EXPLORING
THINK TANKS
DIVERSE ORIGINS,
EVOLVING FUTURES

WORLD SCIENTIFIC SERIES ON THINK TANKS

Series Editors: Enrique Mendizabal
                      Estefania Teran Valdez

Published

Vol. 1   *Exploring Think Tanks: Diverse Origins, Evolving Futures*
         by Enrique Mendizabal

World Scientific Series on Think Tanks – Volume 1

EXPLORING THINK TANKS
DIVERSE ORIGINS, EVOLVING FUTURES

Enrique Mendizabal

On Think Tanks, Lima, Peru

World Scientific

NEW JERSEY · LONDON · SINGAPORE · BEIJING · SHANGHAI · HONG KONG · TAIPEI · CHENNAI · TOKYO

Published by

World Scientific Publishing Co. Pte. Ltd.

5 Toh Tuck Link, Singapore 596224

USA office: 27 Warren Street, Suite 401-402, Hackensack, NJ 07601

UK office: 57 Shelton Street, Covent Garden, London WC2H 9HE

Library of Congress Control Number: 2025041966

British Library Cataloguing-in-Publication Data
A catalogue record for this book is available from the British Library.

World Scientific Series on Think Tanks — Vol. 1
EXPLORING THINK TANKS
Diverse Origins, Evolving Futures

ISBN 978-981-98-1770-2 (hardcover)
ISBN 978-981-98-1771-9 (ebook for institutions)
ISBN 978-981-98-1772-6 (ebook for individuals)

For any available supplementary material, please visit
https://www.worldscientific.com/worldscibooks/10.1142/14432#t=suppl

Desk Editors: Aanand Jayaraman/Venkatesh Sandhya

Typeset by Stallion Press
Email: enquiries@stallionpress.com

About the Author

Enrique Mendizabal is the founder of On Think Tanks (OTT) and has over 20 years of experience in the field of evidence-informed policy, working with and supporting think tanks. Before founding OTT in 2011, he worked for the Overseas Development Institute (ODI), where he headed the Research and Policy in Development program. At ODI, he co-founded the Outcome Mapping Learning Community and began to explore think tanks and their impact. Enrique has co-founded various initiatives aimed at supporting better-informed policymaking, including the Peruvian Alliance for the Use of Evidence, the Latin American Evidence Week, and the Premio PODER al Think Tank Peruano del Año.

Contents

Section 1

Setting the Scene

Chapter 1

Introduction: Think Tanks as a Subject of Study

Think tanks across the world are enjoying the attention of international and local donor communities. Political leaders, governments, and the media have also taken notice. Hundreds of millions of dollars were earmarked to support their development over the last decade, and hundreds more have been spent on them as they are recruited as partners in the fight against global poverty, climate change, democratic backsliding, and the weakening multilateral system — among other causes. The Think Tank Initiative, for example, funded by the William and Flora Hewlett Foundation, the Gates Foundation, the British Department for International Development (now FCDO), the Netherlands Directorate-General for International Cooperation, and the Canadian international Development Research Centre (IDRC), committed over US$100 million in funds to support think tanks in Latin America, Sub-Saharan Africa, and South Asia between 2009 and 2018. The Open Society Foundations ran a similar initiative in support of think tanks in Eastern Europe and the Western Balkans, the Think Tank Fund; other initiatives such as the Australian Government's Knowledge Sector Initiative aimed to revitalise Indonesia's knowledge and the U.K.-funded Zambia Economic Advocacy Programme offered think tanks ample opportunities to access new significant sources of funding.

More recently, the World Bank has launched a series of efforts to support think tanks in Africa, so has the German cooperation agency, GIZ.

This interest from international development funders is matched by domestic political philanthropy that has turned to think tanks in their efforts to influence public policy at best, and capture the state at worst. Most famous is the Heritage Foundation's successful Project 2025, which helped develop and deliver President Trump's agenda in 2025.

Attention and funding, however, do not come without a cost. Think tanks are beginning to face new challenges associated with this newly found popularity.

For six years, since 2004, I worked for an international development think tank in London, the Overseas Development Institute — now ODI Global (ODI). At ODI, I was part of, and later led, the Research and Policy in Development (RAPID) programme. RAPID focused its attention on the study and promotion of the links between research and policy in international development policy and in developing countries. Naturally, think tanks played a key role in our work. Upon leaving ODI in December 2010, I founded On Think Tanks and have focused my own work on the study of think tanks. This book is an attempt to bring together some of what I have learned so far.

Before ODI, back in Peru in the late 1990s and early 2000s, I had begun my professional life working for the Research Centre of the Universidad del Pacífico (Centro de Investigación de la Universidad del Pacífico or CIUP) and so frequently engaged with other think tanks. Only that, back then, we did not call them think tanks. CIUP was a research centre, or *centro de investigación*. And the "think tanks" that we worked with were also research centres, non-governmental organisations (NGOs), or the academic departments of other universities. It is only relatively recently, when funders decided to support think tanks via programmes with "Think Tank" in their names, that many of these organisations made the conscious decision to label themselves as such. The same self-labelling exercise has happened all over the world. And in some cases, funders, the media, and other political actors are relabelling organisations without them knowing or having a say in this process.

Organisations that we might not initially recognise as think tanks in the United Kingdom or the United States, like NGOs or advocacy or research networks, are now presented as think tanks. At ODI, we, too, were quick to do the same. When I joined, the RAPID programme was setting up a network of organisations interested in the study and support of evidence-informed policy, the Evidence-based Policy in Development Network (EBPDN). Although the initial group of members was rather eclectic, including non-governmental organisations, consultancies, and even some civil society networks, we quickly fell into the custom of calling all of them think tanks. Both because ODI was at the time looking for other think tanks in developing countries to partner with and because some of them were calling themselves think tanks in response to donors' interest, we went along with it.

However, when we started to treat think tanks as a focus of study, we found ourselves in a rather interesting and sometimes uncomfortable situation. The sector was terribly diverse, and it was difficult, if not impossible, to come up with a single definition that would suit all the organisational structures out there. On the back of research I had done on policy research networks (Mendizabal, 2006), which suggested that "form should follow function" in the development of networks, we opted to avoid definitions that focused on how the organisations were structured and instead on the functions these organisations were carrying out.

By doing so, it was possible to include a broader range of organisations in our analysis and interactions. We would not have to leave anyone out.

Inclusivity presents new challenges.

In the end, we put organisations as diverse as non-for-profit research consultancies, research departments of universities, research networks or programmes, government research institutes and teams, research departments within NGOs, advocacy organisations, research consultancies, and more traditional independent think tanks into the same bag. This created confusion among our partners in the EBPDN and eventually led to the dissolution of the network.

We were not alone. James McGann's popular yet controversial Global go-to-Think Tanks reports, published until 2020, include research funders such as the Canadian International Development Research Centre, or international campaigning organisations like Amnesty International and Transparency International, among the organisations he labels as think tanks. These are, as discussed in the book, not think tanks.[1]

---

[1]James McGann's annual ranking of think tanks and the reports he and his team at the University of Pennsylvania published adapted the definition of think tanks over the years. From a very U.S.-centric normative definition of what a think tank should be like, the reports increasingly adopted a more relaxed approach, including organisations affiliated to governments, universities, and the private sector. Unfortunately, without rigour in their analysis of the organisations or reflecting on their contexts, the results involve serious mislabelling and misrepresentation of think tanks.

The role that research and analysis played in all these different types of organisations was also varied — with only a few showing sufficient capacity to conduct high-quality research and most either using existing research or conducting only brief analyses or commissioning consultancies to inform their advocacy work.

I now realise that this approach to defining think tanks based on their functions rather than their structure could only take the study and support of think tanks so far. It is neither possible nor useful to treat structurally different organisations in exactly the same way, even if they fulfil similar functions. Doing so led to a number of mistakes in the way that we, at RAPID, engaged with them, the expectations we had for each other, and the delivery of our work (Mendizabal *et al.*, 2011).

Since founding On Think Tanks in 2010, I have had the opportunity to engage with and work with thousands of think tank leaders across the world. The On Think Tanks programme has documented their missions, business models, organisational arrangements, and strategies. We have learned from their successes and failures. OTT has recorded individuals' professional journeys across the think tank world — from the foundation of think tanks to their closure.

The OTT team has witnessed the rise of think tank models in Eastern Europe and the Western Balkans, in Asia and the Middle East. We've witnessed the new struggles that think tanks in Europe are facing, the new models emerging in China and in Saudi Arabia, and the reframing of the "think tank" label.

Ultimately, throughout our work, we have always tried to encourage a more local and flexible definition of think tanks. One that responds to its context because it has emerged from its context.

Studying Think Tanks

Approaching the study of think tanks in contexts where the label is relatively new or little understood poses conceptual, methodological, and operational challenges that need to be addressed. First, there is the problem that there is no standard definition of think tanks — they range from the legal, as defined by national legislation; to the intuitive, like James McGann's "I'll know one when I see one"; and even to the mythical "wizards" according to Fran Luis Kwaku (2005). Some scholars have settled on somewhat sanitised definitions that remove all signs of politics and ideology, while others stress the complex nature of think tanks' private and public relations with political, social, and economic power.

Definitions

The difficulty in defining think tanks is influenced by the fact that the *think tank* label itself cannot be easily adopted and adapted across nations — Germans are uneasy with the idea of armoured vehicles pushing forward policy ideas and so prefer to refer to them as *fabriken* (Braml, 2004). There is also the issue that the term is in English and does not easily translate into other languages — which often means that the Anglicism is used instead. Not everyone agrees that this is a good thing. Fernando Straface, the former Director of the Argentinean Centro de Implementación de Políticas Públicas para la Equidad y el Crecimiento (CIPPEC), argued that using the English term instead of the Spanish alternative "Institutos de Investigación de Políticas Públicas" saves him time. However, he concedes that this can create confusion among outsiders who do not know what the English term refers to: Is it a glass tank (as in fish tank — as in the original concept) or an armoured vehicle?

The definition is further complicated by the roles that think tanks are expected to fulfil in different contexts. For instance, in Saudi Arabia, think tanks are described as "centres for decision-making support" (Alansi, 2021); such an organisation would be considered a policy unit in the British context.

Additionally, the perception of think tanks tends to be rather negative or suspicious: Policy wonks (U.K.), *izquierda caviar* or caviar leftist (Peru), Colombo liberals (Sri Lanka), and liaison lieutenants (Bangladesh) are just a few examples of terms used to poke fun at them.

In some parts of the world, this negative perception has led think tanks to ditch the label. For instance, the label is associated with pro-European and progressive advocacy organisations in parts of the Western Balkans. This makes it hard for some organisations to operate in increasingly hostile environments.

All this makes it hard to define the focus of this study. I myself struggle to explain to my friends and family what a think tank is. I find it easier to call them research consultancies (when talking to a banker or entrepreneur friend) or applied research centres (when speaking with someone more familiar with academic research or public policy).

Perspectives

The second challenge we face in the study of think tanks refers to finding the most appropriate perspective from which to do so, which, given the absence of a clear definition, can be quite a pitfall. This matters because it affects the importance that think tanks are awarded in the policy space: For instance, some perspectives will overestimate while others will play down their influence.

A group of think tank scholars have opted for a focus on organisations that they have themselves pre-identified as a "think tank." This micro-level analysis often leads to both a think tank-centric view of the world and the promotion of a specific think tank ideal (The Brookings Institution, The RAND Corporation, and The Heritage Foundation, for example), even if this ideal bears little resemblance to the actual organisations it purports to describe. Within this same perspective, other approach to the study of think tanks is to focus on one or more of their core functions, the influence on policy being a favourite one among researchers. These approaches attempt to find out how it is that think tanks go about influencing policies, some from a purely exploratory perspective (almost more interested in the evolution of a particular policy discourse than on the specific actions of think tanks) while others adopt a normative (which assumes one type of influencing that all think tanks are expected to follow) or evaluative approach (with a focus on measuring, and, more recently, quantifying, think tanks' influence).

More interesting, in my view, is to study think tanks from the outside. In this book, I consider the elitist, statist, and pluralist perspectives of policymaking because they explain think tanks as part of a wider context and both the consequence of and an influence on political, economic, and social forces.

Operational challenges

How to go about studying think tanks? Given the diversity of definitions and perspectives that exist, we are faced with multiple challenges: Is the existing literature (mainly from a few high-income countries with a long and rich think tank history) relevant for other contexts? What would be the right mix of case studies for a study that attempts to add value to the debate on think tanks across the world, across what Diane Stone calls "think tank traditions"? How many cases are enough? Should we focus our attention on think tanks, countries, or regions? Should we assess all think tanks' functions, or are there any that can be said to be all-encompassing and therefore an entry point into all others?

This book comes at a time when think tanks are experiencing a new wave of interest, and this presents new pressures for what I would hope would be a measured and candid study of these organisations.

On the one hand, this interest has opened up the debate on think tanks and increased the demand for more up-to-date and relevant research on the subject — something I hope this book will respond to. On the other hand, it has led to the formation of a thicker skin among some think tanks that feel that this increased attention means more scrutiny and criticism, as well as a sense of urgency to develop new ground-breaking and practical knowledge on the matter.

This book tries to address some of these challenges. First, I will discuss some of the multiple definitions that are being used in the literature. I will not, at this stage, present a final definition but rather develop a working definition that draws from several of these different approaches. In reviewing the definitions, I have found that important questions arise about the nature of think tanks in developing countries and the possible consequences that preferring certain definitions over others may have.

I then attempt to outline some of the main schools of thought within which the study of think tanks is taking place. These schools of thought will offer a sense of the different perspectives from which think tanks can be studied. Their differences reflect the nature of the political context in which think tanks exist and so should provide guidance for future research. They also provide insights into the roles that elites and the state play and debunk some myths about the importance of democracy for their formation. Rather interesting findings about the positive roles that non-democratic regimes and institutions such as the Church can have are presented.

As with most studies of think tanks nowadays, special attention is given to think tanks' policy-influencing function. Within this analysis, I briefly consider and discuss the strengths and weaknesses of different analytical frameworks to study think tanks and their development more generally and to assess their influence on policy more specifically.

Other topics of interest have arisen as well. For example, the nature of national political systems and the nature of funding for think tanks can help explain the size of the think tank community in any given country.

Also interesting is that the type of funding think tanks receive, rather than its quantity, emerges as a critical factor for our understanding of think tank differences. Powerful ideas, such as the belief in the positive contribution that science can have on policymaking, also appear to explain many of the key characteristics of think tanks and can therefore provide an explanation for their diversity. The specific roles of different funders in the setting up and development of think tanks also seem to affect their strategic choices, for instance, the degree of openness of their communications work or their ideological discourses. These are just some of the ideas emerging from the analysis that are presented in this book.

I have made use of cases and examples from think tanks to address these questions. Some of these have been drawn from my consulting work for think tanks over the last 20 years. I have complemented this with countless substantive visits and engagement with think tanks in different contexts, including the U.S., Argentina, Peru, Ecuador, El Salvador, Mexico, Tunisia, Saudi Arabia, Zambia, Indonesia, Malaysia, Serbia, Georgia, Hungary, Germany, the United Kingdom, China, Vietnam, Bangladesh, and Myanmar, and have provided new insights from the literature gathered over the course of writing this book.

The On Think Tanks blog, interviews, and podcasts, produced in collaboration with think tankers from across the world, have served as an invaluable source of inspiration and evidence. More recently, OTT's Open Think Tank Directory and the annual Think Tanks State of the Sector Reports offer a rich set of data and analysis to better understand the field.

These examples, the data, and the new literature studies provide us with an opportunity to explore new emerging lessons on think tank development of immediate relevance to think tanks themselves and to the organisations and people that support them.

Approaching the Literature

Undertaking a literature review on think tanks presented both a challenge and a great opportunity to break some of the consensus that had hitherto limited the study and work on think tanks. Rather quickly, the search for research on think tanks led me to debates on the roles of technocrats, ideologues, public intellectuals, and various other interest groups.

Technocrats led to experts and a discussion over whether think tanks may be in fact contributing to the weakening of democratic debate (Mendizabal, 2022), an issue that had been consistently taken up by British journalists George Monbiot and Adam Curtis and that has become an issue of great interest for political and economic commentators as technocrats gained the upper hand over politicians in the aftermath of the Eurozone crisis.

The first Trump administration and the run-up to his second election renewed interest in think tanks' transparency and the undue influence of domestic and foreign interests. The work of Michael Schafer for Politico (Michael, 2025) and Ben Freeman and Nick Cleveland-Stout for the Quincy Institute (Freeman and Cleveland-Stout, 2025) over the last 5 years picked up on work by Eric Lipton a decade ago on the undue influence of foreign agents (Lipton *et al.*, 2024).

The role of ideologues led to the role of political parties as key users and promoters of policy research and think tanks in particular. In recent times, ideologues and political parties have been seen as contributing to a growing sense of scepticism on evidence and think tanks.

The literature on public intellectuals was particularly interesting, and I am hopeful that this review will award it the attention it deserves for the study of think tanks. But most fascinating was the debate on the changing nature of policymaking and the increasing privatisation of public decisions. This issue of the privatisation of public decisions is particularly relevant as more and more decisions are increasingly outsourced to automated decision-making and Artificial Intelligence.

This book could have ventured into the worlds of specific disciplines: economics, foreign policy, military and defence, and environmental sciences, for example. Instead, I opted to remain *neutral* to the subject matter that think tanks deal with. However, this neutrality is quite arbitrary and difficult to maintain, and so, in cases when it is clearly relevant to explain think tanks' formation and development, I will attempt to bring the content back into the analysis.

I have also resisted the temptation to treat "international development" as a research subject matter of relevance for think tanks in low- and middle-income countries or the Global South, except when reflecting on international development think tanks based in donor countries or the effect that international development policy from donors and bilateral and multilateral agencies has on think tanks.

There is a reason for this. The Zambian Institute for Policy Analysis and Research (ZIPAR) or Ecuador's Grupo FARO, mentioned in this book as examples, are, in my view, closer to the Institute of Fiscal Studies (IFS) or the Institute of Public Policy Research (IPPR) in the United Kingdom than they are to ODI Global or the Institute for Development Studies (IDS).

The former, as in the case of their peers in developing countries, focus on mainstream politics, while the latter deal with the marginal politics of international development, which have, at least in the United Kingdom, become almost entirely depoliticised.[2] Unfortunately, given that much of the funding for southern-based think tanks comes from bilateral, multilateral, and northern-based foundations, the international development industry, in this case, does play a disruptive role in think tank communities in the Global South (Mendizabal, 2024).

Finally, I am sure that there are many more cases, studies, and entire bodies of knowledge of relevance that the reader will no doubt know about and that I have missed. I hope to have a chance to use some of these in future studies and the discussion that this book, I expect, will trigger.

---

[2]By this, I mean that for a long time international development policy appeared to be developed and implemented outside the traditional domains of political ideology: No major differences could be identified across the major three political parties in the United Kingdom. Even now when new political players advocate against Aid such as Reform and a move by the Labour government to cut international development funding, the policies of international development do not concern most of the public.

Chapter 2

Think What? Attempts to Define and Describe Think Tanks

So, what is a think tank? This chapter is organised across four types of answers to this question: normative definitions based on an idealised idea of modern think tanks, descriptions based on think tank characteristics, relational definitions, and functional descriptions. These are all easily identifiable in the literature and have also emerged from my conversations with staff from several think tanks over the course of my work. I had considered dodging this question altogether and arguing that, as James McGann is famously quoted as saying, I'll know one when I see it. However, a comment by Fred Carden, formerly at IDRC, encouraged me to tackle it. Quite clearly, even the most flexible approaches cannot escape the fact that any meaningful discussion about think tanks needs to start from somewhere, even if that is a limited subset of organisations that fit a particular definition. If, however, we are willing to let the boundaries (Mendizabal, 2014) of this definition expand as we explore new contexts, cases, and perspectives, then we will be on the right track.

Another reason to tackle this and attempt to define the boundaries of the definition is that, throughout my work with think tanks, I have noted the influence that a few think tank have had on new think tanks. I was of the view that having the Brookings Institution as the model think tank was an undesirable and even counterproductive idea that failed to recognise the diversity of think tanks across the rest of the world. However, it is clear from talking to think tank directors and staff, who are in many cases also their founders, that several think tanks have been set up with a Brookings stereotype in mind. So it is, even if operationally irrelevant to their realities, relevant to their origins, histories, and motivations. This does not mean, however, that we should attempt to compare them, like for like, with Brookings (Mendizabal, 2019). A more nuanced reflection is necessary and possible.

This book refers to several cases of think tanks that were set up or managed by *returnees* in Sri Lanka, Peru, Serbia, Germany, China, Malaysia, and Zambia, among others.

Experienced and entrepreneurial alumni of Western universities and think tanks who, having been exposed to and inspired (or at least informed) by the Anglo-American think tank model, saw a gap and an opportunity to replicate them in their own countries. Foreign donors have played similar roles in founding and funding think tanks with the expectation that they would naturally develop into Southern versions of *Brookings*, *Chatham Houses*, and *Overseas Development Institutes*. The African Capacity Building Foundation (ACBF), for example, has been a key proponent of the university-without-students model that is supposed to characterise the Brookings ideal. As a consequence, ACBF's agency has had significant effects on the think tank community in Africa, even if none of the think tanks it supported have ever lived up to the ideal — nor could they.

When the founding executive director of the Red de Desarrollo (REDES) was working on the new think tank's original strategy in 2019, he explicitly referred to the expectation that the new think tank would eventually develop into "a Peruvian Brookings." He was referring to Brookings' reputation and permanence in the U.S. policy landscape rather than to its business model, but its influence and significance are clear.

Across the world, many think tanks have been effective in moulding themselves after their peers elsewhere. It is not impossible to identify well-funded economic, foreign policy, security, and military think tanks across Latin America, South Asia, and Southeast Asia that are indistinguishable from their Anglo-American inspirations or from the other think tanks that inhabit their policy communities, except maybe in their global profiles and influence. Take the Gateway House in India, a foreign policy think tank founded 2009 that looks and feels like any other foreign policy think tank in the West; Brazil's Fundação Getulio Vargas is bigger and better funded than most U.S. based think tanks; and Zambia's Policy Research and Monitoring Centre (PRMC) is a political think tank like the ones found in Britain.

Even the promoters of think tanks in contexts as different as Saudi Arabia or China have Anglo-American models in mind. More recently, think tanks across the Gulf region have invested in foreign think tank experts, study trips, and partnerships with U.S.- and U.K.-based peers. The demand for ideas and experience bears no relation to the impossibly more different realities.

Tackling the question of how to define think tanks, therefore, presents an excellent opportunity to study the very nature of the diversity found among these organisations across the world. And attempting to draw boundaries around certain organisational and functional characteristics will inevitably force us to think about the purpose, the *telos*, of think tanks, and, in a modest Aristotelian sort of way, determine who deserves the label.

Written into Law: Normative Definitions

At the very basic level, a think tank (and since the label emerged there) can be defined by what the United States' legal code says: 501 (c)(3) organisations are non-for-profit, non-partisan, and organised for educational, religious, charitable, and scientific purposes (Harvard Law Review, 2002). They cannot, for example, support a particular candidate or undertake lobbying activities. Americans may have named them, but Smith (1991) says that think tanks in the United States were inspired by the British experience from the mid-1800s. The Open Think Tank Directory identifies the oldest think tank in existence as the Royal Society of Arts, Manufacturing and Commerce (The RSA).

Many think tank scholars have adopted this basic definition to develop "working definitions" that stress similar qualities. This has become known as the Anglo-American lens by some who often criticise it for not being relevant to the characteristics of developing countries. But is it?

Several authors have used it to the explore the ways in which think tanks can be defined and classified in both developed and developing contexts (see Garce and Una, 2007, Abelson, 2006, 2009, Rich, 2006, Smith, 1991, Weaver, 1989, McGann and Weaver, 2002, Stone and Denham, 2004). As a consequence, the official definitions of think tanks used by some think tank funders are still based on their organisational setup and identify them as non-profit, independent of the state, and dedicated to communicating research findings to policymakers. Not surprisingly then, think tanks with these characteristics are more often found in the United States, with a much smaller number in Canada, the United Kingdom, Australia, and New Zealand. Many think tanks in the Global South fall wayward of this definition: for example, those linked to the state or universities but so would think tanks in Germany, which are significantly dependent on public resources, often have a semi-public status, and are often formally affiliated with political parties, or think tanks in China or Vietnam, which are part of the regime's policymaking system.

This narrow definition also leads to the assumption that think tanks are entirely new phenomena. As if, somehow, the role of independent policy advice did not exist before the label was invented in Britain and the United States, or, indeed, before it became fashionable among international development donors to support them and many research centres, non-governmental organisations, and even consultancies started to call themselves think tanks.

There are valid criticisms. Tom Medvetz (2008) has argued that this definition is particularly limiting because of the following reasons:

(1) It privileges the traditions of think tanks in the United States and Britain over all others.
(2) It leaves out many present-day examples that do not fit with the definition: for example, corporatist think tanks in Japan, publicly funded or government think tanks in Vietnam, China, Saudi Arabia, and Germany (even in the United States, think tanks like RAND and the Urban Institute are federally funded organisations), university-based think tanks across Latin America and Africa, and partisan think tanks in Chile, Uruguay, the United Kingdom, the United States, Germany, etc.
(3) It robs the concept of think tanks of historical depth as it forgets that their origins can often be found in the same institutions they are now supposed to be independent of.
(4) Most significantly, it fails to recognise the importance of the concept itself and the political nature of the choice of the label.

This last point is particularly worth noting. Think tanks are not specifically mentioned in legislation. They are just one type of organisation: for instance, an NGO that focuses on research rather than service delivery. Therefore, what is and is not a think tank among all civil society organisations is more likely to be defined in a self-labelling exercise. This is quite important because legislation to encourage or control the formation and development of civil society organisations, as we shall see in Chapter 7, is a powerful means of control of the roles that think tank can play, and also because the influence of external players, particularly by means of their funding practices, can lead to the proliferation of label-only think tanks.

This definition poses several questions. Is it good enough? Is it acceptable to allow organisations to label themselves as think tanks and take their word for it, as long as they fit some legal definition?

As more funds have become available, think tank organisations that may in the past not have considered themselves as such, now do. NGOs with little research or analytical capacity can (and do) claim that their work with grassroots, for example, provides them with "evidence" and by using that evidence in their advocacy work, they are, in fact, undertaking the roles associated with think tanks. Some for-profit consultancies argue that by publishing their "knowledge," drawn from their experts and the lessons learned in the programmes and projects they implement for their clients, they are also fulfilling think tank functions. Academic departments, media outlets, and policy analysis bodies within political parties or government departments are all claiming the same: We are all think tanks!

But are they? And should they be labelled so? Is it in their and the policy research community's best interest that they do so? Should being a think tank imply the acceptance of certain responsibilities that are different from those of other civil society organisations? Being not-for-profit seems a perfectly reasonable one, but is it always possible, for instance, in contexts where civil society is generally under threat — as is increasingly the case (Mendizabal, 2025)?

Surely, a condition that nobody could argue against is being autonomous, to develop and pursue the ideas one wishes to. But what if the centre is associated with another institution? Is it no longer a think tank if it is part of a ministry, linked to a private corporation, or a university? What if it has a stated and identifiable ideology? What about the type of funding it receives? Can an organisation be a think tank if it works on a project-by-project basis, de-facto operating as a not-for-profit consultancy, or should it be entirely disassociated from any commercial activity?

And what about its relationship with its main public? Can a think tank take money from the very people or organisations it seeks to influence? What should be the terms of this transaction? When does a think tank cease to be one to become a consultancy, a service provider, or even an advocate or lobbyist for someone else's ideas and policies?

It should be clear by now that the normative definition cannot answer all these questions. This limitation makes it hard to define the boundaries of the label.

Legislation is not always the driver

The challenge with the normative definition is that think tanks have not just emerged from within the legal framework; instead, they have been innovations typically inspired by external ideas and individuals.

I am not aware of any legal code that includes a definition of think tanks. Take the origin of modern think tanks in the United States, for example: James Smith (1991) calculates their origin to a meeting of one hundred people including writers, journalists, educators, scientists, and government officials at the Massachusetts State House in Boston in October 1865, who came together to explore ideas and alternatives to improve the economic and social wellbeing of the States. This led to the formation of the American Association for the Promotion of Social Science — later known as the American Social Science Association (American Association for the Promotion of Social Science, 1866). In turn, they had based their organisation on a similar British Association founded in 1857 (Smith, 1991).

A similar story in which institutions are imported and adapted is found in the foundation of the Sociedad Académica de Amantes del País, established in Peru in 1790. It, in turn, drew inspiration from the Royal Basque Society of Friends of the Country established in Spain in 1773. More than 200 years ago, the pioneers of modern think tanks were already innovating with models learned from elsewhere and, most crucially, challenging their own societies' status quo.

In China, the story goes that Chairman Mao instructed his premier to establish the first foreign policy think tank concerned by the Party's failure to predict political unrest in Europe and the Soviet Union.

The models we see today across the world have emerged as innovations, big and small, and have piled on top of each other. Hence, expectations of what is and what is not a think tank have always played catch-up with what is, in essence, a fluid idea that is always on the lookout for new models in which to find new expression and meaning.

The normative definition is therefore not flexible enough to address the dynamism that is historically hardwired into think tanks' DNA.

Luckily, there are plenty of other approaches from which to draw a more nuanced explanation of what may be a think tank.

Describing Think Tanks: Categories and Types

The literature provides a rich number of ways to characterise or differentiate think tanks, and this is something think tanks like to do themselves. For example, Reinikie (quoting Smith) suggested the following criteria: their sources of finance, the constituencies they choose to serve, the balance they strike between research and advocacy, the breadth of the policy questions they address, the academic eminence and practical policy experience of their staff, and their ideological orientations (1996, p. 29).

Each poses interesting questions about the nature of think tanks, and I will attempt to address some in this section.

Think tank inequality: Size and scope

A straightforward and obvious way of classifying think tanks is to look at their size. Murray Weidenbaum used basic income and assets to divide think tanks according to the size of Washington, D.C. (2009). He found that most are quite small but the perception of the contrary is explained by the fact that the top 5 think tanks took US$139 million out of US$411 million total income in 2005.

These budgets have grown significantly over the years. In their annual reports, only the American Enterprise Institute (2021) and the CATO Institute (2022) reported budgets of US$49 million and US$39 million, respectively.

This diversity is also found among think tanks in developing countries. Among the Think Tank Initiative's grantees, for example, there were centres with only a few members of staff and others with over 100 researchers.

Large mega think tanks, like those set up in China in the 2010s (Zufeng, 2009) or the foundation of publicly funded Brazilian think tanks, are not common. More common are medium-sized organisations of between 10 and 30 staff working on a range of specific policy issues, which fluctuate based on the interests and concerns of their funders.

Domestically funded think tank communities have tended to be more flexible, with periods of growth followed by periods of downsizing. The U.S., Britain, Malaysia, and Chile, for instance, often see the rise of small politically focused think tanks before or after electoral races. Only some survive the transition of their founding leaders into government positions.

Internationally funded think tank communities, as in most of Africa, Latin America, South Asia, and Eastern Europe and the Western Balkans, on the other hand, have seen more continuous periods of growth fuelled by a growing international development budget. The U.S. Government cuts of 2025 have, however, triggered a rapid retrenchment of funding for think tanks.

Smaller think tanks inevitably develop a focus on one or two policy issues, while the larger ones may be able to cover a greater range of subjects. Weidenbaum links the size and the thematic focus of think tanks in his classification, differentiating between: large and diversified, large and specialised, and small and specialised. Andrew Rich instead classifies think tanks by their scope alone: full-service, multi-issue, and single-issue think tanks (Rich, 2006).

Ray Stryuk, mainly looking at think tanks in Eastern Europe, added historical evolution or progress (Struyk, 2006) to the analysis of size. In his categorisation, think tanks can belong to three groups depending on their stage of development: first stage (small organisations with fewer than 10 staff); second stage (larger with 5–10 or more and managing more projects and with greater specialisation in their assignments and more opportunities to influence policy); and third stage (think tanks which are much larger and complex organisations).

This is a categorisation more commonly used by think tanks in Eastern Europe than in other regions.

What defines the size and scope of think tanks? Two key drivers of growth can be identified, but these are entirely external to the think tanks themselves, and often work against each other: donor funding and human resources. These two factors play out differently depending on the context.

In some countries, donor funding has been increasing as research funders narrow their focus on the poorest countries and set up their own initiatives to develop the capacity of think tanks. This, of course, means that some regions and countries, particularly middle-income economies, have seen funding reduced.

The better-funded think tanks in these poorer countries are turning to their own labour markets for new recruits but find it extremely difficult to fill the new posts. This is the case of think tanks in places as different as Paraguay, Zambia, and Indonesia. Across Sub-Saharan Africa, think tank directors and experts that I have worked with have identified staffing as a key challenge. Similarly, two informal surveys of think tanks in Latin America conducted by both the Think Tank Initiative and the Latin American Executive Directors (DEAL) community of practice found that staffing concerns were among the top challenges faced by think tanks in the region. This concern was echoed by think tanks at the 2025 Central Africa Think Tank Forum and On Think Tanks 2024 and 2025 State of the Sector Reports.

More recently, global foundations have turned their attention towards localising their agendas and work. But in 2024, the director of a leading South African think tank argued that they had struggled to hire economists and explained this was partly due to the increasing competition from foundations.

With few experienced researchers to draw from, some think tanks have to offer rather large salaries and pay packages to attract and retain new staff. In some cases, particularly in countries where international development funding is more readily available, this is possible; it has, however, the undesirable effect of pushing salaries to unsustainable levels since the supply of capable researchers does not meet this increased demand. The sudden cuts in funding from USAID have left many of these think tanks vulnerable.

Where funds are still limited, such as in middle-income countries, the pay packages that think tanks can offer are never attractive enough to tempt the most competent researchers away from better-paid and more interesting opportunities in the private or public sectors — or the safety of the university environment.

The existence or absence of political competition may also be a factor. In the United Kingdom, ODI Global grew every year from about 100 staff in 2004 to close to 200 in 2012. Its budget increased accordingly. Other British centres focused on domestic politics, however, saw their size fluctuate with political and funding cycles. Funded by an aid budget that has seen a consistent increase, ODI Global was able to avoid the ebb and flow that is the norm among its domestically focused peers. Changes to U.K. international development and policy priorities, among other factors, led the centre to a significant restructuring and downsizing process.

Strategy: Funding, Business Model, Staffing, Objectives

The size and scope of think tanks can then be limited by factors that have little to do with think tanks' strategic decisions. Another way of classifying think tanks is to focus on their own strategies. Are there any choices that would define or exclude an organisation as a think tank?

Funding

The source of funding is particularly important to describe think tanks because it can help create nuance in the understanding of the key characteristics, such as whether they are non-profit or independent. Weidenbaum developed Table 1 to compare the top 5 D.C.-based think tanks in the United States according to the source of their funding (2009, p. 39).

Back in 2005, the Brookings Institution clearly depended on its endowment, the American Enterprise Institute on donations from individuals and corporations as well as income-generating activities, CSIS almost entirely on individual donors, Cato on corporations and foundations, and Heritage on individuals and foundations.

Unlike what one would find in Germany and many developing countries such as China, Vietnam, or Brazil, government funding is negligible for the largest D.C.-based think tanks (Braml, 2004). The cuts in federal funding for civil society organisations introduced by the second Trump administration in 2025 did not have a significant impact on think tanks. A senior member of staff at Brookings, for instance, explained that even

Table 1.   Funding sources of the largest D.C.-based U.S. think tanks (2005), in percentage of total income.

	Individuals	Corporations	Foundations	Government	Endowment	Sales and Conferences
AEI	37	21	16	0	0	26
Brookings	20	17	19	2	25	17
CSIS	83	2	11	0	0	4
Cato	12	28	35	9	5	11
Heritage	59	5	24	0	8	4

after 4 years of a Democratic Party administration, the think tank remains largely independent from government funding.

However, the RAND Corporation and the Urban Institute, which were not included in Weidenbaum's study, have historically received government funding, but this reflects their origins and business models.

Smaller think tanks across the U.S. have also been affected.

Funding in low- and middle-income countries for individual think tanks is less easy to assess, but interviews with think tank directors and staff suggest that grants from individuals, corporations, or foundations (and certainly endowment) are hard to come by. By far the most common type of funding reported is project funding from international development agencies, long, medium, and short term (On Think Tanks, 2024).

Over the last decade, even if global foundations have publicly championed the benefits and need to move towards long-term and core funding for civil society organisations, many directors express concerns that their funders, who in the past might have provided longer-term funding (through projects or even core funding), now prefer to rely on shorter-term consultancy contracts in their dealings with them.

This responds to several overlapping agendas in the international development sector. First, the value-for-money agenda places, in my view, unnecessary and unrealistic accountability pressures and success expectations on think tanks. Second, a mission-driven agenda has moved many funders to set and demand increasingly unrealistic short-term goals for think tanks.

This unequal pattern of support is reported in the 2024 OTT State of the Sector Report, which found that while think tanks in high-income countries receive a significant portion of their income from programmatic and core funding, think tanks in low- and middle-income countries are more dependent on project funding (On Think Tanks, 2024).

This dependence on short-term and foreign funding in low- and middle-income countries imposes significant constraints on the development of local think tank communities. The 2025 USAID cuts in funding have affected think tanks as well as other civil society organisations across the Global South.

A 2025 study by Accountabilitylab Pakistan states that almost 50% of the civil society organisations surveyed reported that the funding freeze was a "financial catastrophe" and their hope for substituting this loss of funding was low: "only 9% organisations are expecting to substitute more than half of the USG funding gap."

Alternative private domestic funding (public or corporate) of the kind that would help think tanks create opportunities for independent thinking is only readily available in more developed middle-income economies and political systems (in Chile, Brazil, Argentina, Mexico, and Colombia, for example; in South Africa; in India; in Indonesia, Vietnam, Malaysia, and China). Where public domestic funding is more common, however, this tends to be publicly provided to think tanks that exist either within the state apparatus or within its direct sphere of influence (Hayter and Makokha, 2024).

As a consequence of this dichotomy, between domestic and international sources — and the importance of foreign funding, even in a country like India — Srivastava (2012) suggested the following classification of think tanks in South Asia:

- Government-funded but autonomous think tanks.
- Foreign-funded think tanks.

This difference is important. Srivastava recognised that even government-funded think tanks in the region are not often able to meet all their funding needs and thus rely on additional foreign funding. By and large, though, they tend to focus their research agendas around issues of economic development within the prevailing official narrative. Foreign-funded think tanks, on the other hand, exhibit a mostly project- or contract-based model and focus their research on issues of human rights, governance, and democratisation, for example, again, within the prevailing international development narrative.

His study, however, did not cover a number of Indian foreign policy and security think tanks that are funded privately. For example, the Gateway House is a foreign policy think tank funded by the Mahindra Group, T.V. Mohandas Pai (Chairman of Manipal Universal Learning), Suzlon Energy, Rakesh Jhunjhunwala (the "Warren Buffett of India"), and the TVS Motor Company.

Considering these different funding options, it would be possible to describe think tanks' funding as shown in Table 2.

Business models

To effectively assess how funding affects think tanks, we must first consider their business models (Ralphs, 2016). Think tank scholars seem to agree that a useful way of describing think tanks' business models is to

Table 2.   Sources of funding and income streams for think tanks.

	Endowment	Core Organisational Funding	Independent Initiatives or Programmes (Medium/Long Term)	Projects or Contracts (Medium/Short Term)	Good and Services
Government funding through:					
Private sector funding (including individuals, foundations, and corporations) through:					
Foreign funding through:					

consider both the balance between academic research and advocacy work and the manner in which they earn their income (Abelson, 2006, 2009; Belletini, 2007; Ricci, 1993; Rich, 2006; Reinicke, 1996; Smith, 1991; Weaver, 1989; Braml, 2004; Srivastava, 2012).

Mercedes Botto (2011) and Norma Correa took a similar approach in studies of Latin American and Peruvian think tanks, respectively. In the studies that they carried out for a scoping exercise for IDRC's Think Tank Initiative, and that Mercedes Botto then revised for publication, they took into consideration the origins of think tanks to describe their business models: academic, corporate or entrepreneurial, NGO, and partisan for Latin America. Norma Correa also includes a fifth category that describes think tanks linked to grassroots, unions, or public bodies.

Academic think tanks

Academic think tanks are typically referred to as universities without students. For example, the Brookings Institution, the American Enterprise Institute, Chatham House, and the Center for Global Development could also be called independent or autonomous think tanks. In other words, these are think tanks that are relatively free to develop and pursue any ideas and policy recommendations that they wish. But it would be a mistake to assume that these think tanks do not have ideological or partisan biases. Their independence is precisely demonstrated by their capacity to assert their values and affiliations.

This assertion of autonomy is possible because many developed country think tanks in this category tend to enjoy hefty endowments or stable income streams, while their academic counterparts in developing countries can only rely on the cross-subsidising capacity of the universities or other organisations to which they are commonly formally affiliated. By the same token, it would be a mistake to assume that because a research centre is affiliated with a university, a government, or a political party, it may lack the autonomy to develop its own research agenda. However, the more conditional nature of the funding relationship can raise some concerns.

In fact, the closest examples of these universities without students in the developing world may be more easily found in international research organisations, such as the UN Economic Commission for Latin America and the Caribbean (based in Chile), the UN Economic Commission for Africa (based in Ethiopia), the Center for International Forestry Research (based on Indonesia), the International Food Policy Research Institute (with several regional offices), and other similar centres. At the national level, some think tanks linked to universities are able to draw on reliable and long-term funding that allows their researchers to embark on some truly independent work. CIUP in Peru, for example, expects and funds its researchers to conduct what it terms as internal research projects on an issue of academic and policy interest to them. Ironically, this administrative dependence makes the centre rather more intellectually autonomous than some of its administratively independent comparators. Others, such as the Zambia Institute for Policy Analysis and Research (ZIPAR), were set up with sufficient funds to cover all their research staff for at least a couple of years and therefore enjoyed a de-facto intellectual and operational autonomy to pursue any particular research agenda (although in this case, it is partly defined by specific demands from the government, which is a core founding member). Other interesting examples of this category of think tanks can be found in the state-funded Chinese and Vietnamese Academies of Social Science, although the latter has been under increasing pressure to look elsewhere for funding.

The close relationship with universities poses an important question for the definition of think tanks with links to other organisations. When is a think tank better described as an academic research centre, the research department of a public body or development agency, or the strategic or policy unit of a political party? Is it possible that the concept of independence or autonomy ought to extend beyond the choice of research or policy argument to include administrative procedures and policies? For instance, could a research centre be labelled as a think tank if its brand was indistinguishable from that of the university? This would certainly signal a lack of operational as well as intellectual autonomy. How different would it be from the research departments of companies such as Citibank, McKinsey, or IBM?

Entrepreneurial think tanks

Another important category identified by Correa and Botto is contract or entrepreneurial think tanks. These are highly dependent on projects from private and, mostly in developing countries, public clients (including international financial institutions, development banks, and public and private donors). They are best exemplified by examples such as ODI Global in the United Kingdom, the Center for Poverty Analysis in Sri Lanka, GRADE in Peru, the SMERU Research Institute in Indonesia, and CIPPEC in Argentina.

Most international development think tanks – particularly those in low- and middle-income countries – operate through a series of short- and medium-term contracts, effectively functioning as "research consultants." This model is perhaps the most common in the sector. There are, however, a few exceptions. The Center for Global Development (CGD) in the United States and the Center for International Governance Innovation (CIGI) in Canada both benefited from the generosity of dot-com philanthropists. The Australian National University's Development Policy Centre has secured long-term grants from both public and private sources. And the German Institute of Development and Sustainability (DIE) is publicly funded. But beyond these cases, it has been difficult for most think tanks to attract individuals or foundations willing to provide long-term core funding or endowments.

Instead, they depend on projects or medium-term institutional funding, often in the form of accountable grants or contracts.

The challenge with this category of think tanks is that the contract model is hardly ever the only one in operation. All the think tanks mentioned here also benefit from other forms for funding and are therefore able to manage alternative business models: CIPPEC, for example, launched an independent research and advocacy programme to address the 2011 national elections using funds generated from fundraising efforts targeted at individuals and corporations. And in fact, several others, GRADE, for example, use these contracts to pursue academic interests and publications.

For an important number of contract think tanks, however, more common is the use of funds for specific programmes or projects that are sometimes indistinguishable from those being implemented by for-profit consultancies or NGOs. What is the line then between a consultancy and a think tank? Should there be a balance that must be struck between independent research and contract work to deserve the label? There may be certain contractual obligations that, if accepted, would help draw the line between one and the other type of organisation. For example, is it acceptable for a think tank to accept a contract that has a clear indication of expected findings, outputs, and outcomes? Would this not be more akin to a contract for services than a research grant?

Similarly, the relationship between the funders and the think tanks emerges as a potential way of defining the boundary. Would we call an organisation a think tank if it were routinely sub-contracted by for-profit consultancies or corporations, themselves providing a specific service to a third party? Value Added Tax regulation in the United Kingdom, for example, drove ODI Global to set up a limited company to manage its *commercial* activities: workshops, book sales, and contracts with for-profit clients. Would this alone challenge its status as a think tank? Probably not, but the boundary move nonetheless.

NGO, advocacy, or overtly political think tanks

A third category of think tanks, found by Correa and Botto, is NGO or advocacy think tanks. These think tanks are more focused on the policy process, are typically driven by advocacy and campaigning imperatives, and are comparable, albeit not ideologically nor structurally, to such iconic organisations as the Heritage Foundation and the Cato Institute in the United States. In developing countries, they have mostly emerged out of sections of the organised civil society concerned with human rights or local projects or networks of international NGOs. Some scholars also include the so-called vanity think tanks (*patronised* by an individual or organisation) that have gained popularity in China since the 1980s (Li, 2009) in this category.

These sub-categories of advocacy think tanks have been historically linked to the rise of political parties and think tanks in Colombia (Londoño, 2009) and have a much more specific and narrow audience in mind. Naturally, of course, they are linked to the rise of lobbies and other interest groups as they constitute a clear opportunity to seek influence and power. Unlike the academic and contract centres, advocacy think tanks are typically closer to advocacy NGOs or campaigning publications such as magazines or newspapers with clear ideological or partisan objectives. And as a consequence, they are more comfortable working with the media.

Nonetheless, this advocacy mode of work is growing even among the academic and contract think tanks. The changing nature of funding towards more explicitly political objectives means that many think tanks are under increasing pressure not just to communicate their research findings but to go beyond that and actively advocate for particular policy changes. An important question to ask in relation to the search for the boundary between think tanks and other organisations is whether the arguments being put forward are the think tanks' own and drawn from independent research, or if they have been adopted and are entirely based on values or the interests of a political, economic, or social group. The latter may be better described as an interest group or even a lobby.

Partisan think tanks

All think tanks are political. But some think tanks are partisan. Partisan think tanks are explicitly aligned with a political party. Leandro Echt has studied partisan think tanks in Argentina and defined partisan think tanks as (...) "organizations dedicated to the generation of ideas related to policy and technical advice directed to the interests of a political party to which they explicitly adhere and with which they establish direct collaboration strategies" (Echt, 2019, p. 11).

These think tanks are not common, partly due to the limits that legislation imposes on civil society organisations seeking a not-for-profit status, and partly due to the risks involved in aligning think tanks' interest too closely to those of political parties (Tanaka *et al.*, 2009). They are common in Europe, where political foundations tend to play a think tank role for their parties, as shown by the following examples:

- Konrad Adenauer Stiftung (KAS) is affiliated with the Christian Democratic Union (CDU).
- Friedrich-Ebert-Stiftung (FES) is affiliated with the Social Democratic Party (SPD).
- Heinrich Böll Stiftung (HBS) is affiliated with the Alliance 90/The Greens.

Their status as think tanks is debatable, but they certainly play a partisan think tank role. Political party think tanks are common in Malaysia — where many political leaders have founded think tanks in support of their political enterprises — and single-party systems such as China and Vietnam.

Most think tanks that we may typically associate with this category, the Heritage Foundation in the U.S. or IPPR in the U.K., given how publicly aligned they often are with the Republican and Labour parties of their own countries, respectively, in fact, go to great lengths to demonstrate their differences.

During the 2024 U.S. presidential election, the Heritage Foundation's Project 2025 closely aligned with the Republican campaign. To avoid breaking civil society and tax legislation, it has had to employ a combination of tactics including avoiding electioneering and direct endorsements, focusing on technical issues and framing its work as educational, and using separate entities for different types of activities (e.g., Heritage Action for America, a 501(c)(4), can legally engage in lobbying and more overt political advocacy).

Embedded or inside-track think tanks

Over the last decade, a new model of think tank has emerged as a mechanism to accelerate the use of evidence in policymaking. Often called policy labs, these are organisations, teams, or projects that play several of the functions of think tanks while embedded in a policymaking entity such as a ministry, parliament, or international organisation.

In a study on policy labs by Emily Hayter and Marcela Morales, policy labs are loosely defined as organisations that are close or embedded in government, focus on the use of evidence by decision-makers, and narrowly ground their recommendations on evidence and, in particular, experimentation (Hayter and Morales, 2023).

As the model has developed, it has adopted different forms and now includes initiatives that are entirely dependent on their hosts (e.g., the MineduLab at the Ministry of Education in Peru), semi-autonomous (e.g., the Education Endowment Foundation in the United Kingdom), or independent (e.g., SUMMA — Laboratorio de Investigación e Inovación en Educación para LAC).

At the core of the model is the imperative to maintain a direct line of communication and engagement with decision-makers, often in the technocratic side of policymaking. This differentiates them from partisan think tanks.

Grassroot think tanks

A new set of think tanks has emerged over the last decade. Grassroots or volunteer think tanks, like foraus in Switzerland, or student-led think tanks have been successful in mobilising individuals interested in playing a role in public life through one or more of the functions of think tanks.

These think tanks are characterised by a reliance on volunteers across the organisation: governance, management, research, communications, and fundraising are largely dependent on volunteers.

Foraus combines this with a few paid positions which require full-time commitments. On the other hand, student-led think tanks like STEAR or the Warwick Think Tank have successfully developed and sustained their organisations without any paid positions.

The volunteer element provides these organisations with a degree of connection to their communities that other think tanks struggle to achieve. An important factor in attracting participation is the more explicit opportunity to develop future competencies and skills that these think tanks offer.

Student-led think tanks, in particular, are a powerful training ground for future think tankers.

What affects the business model category?

As briefly discussed for each of the business model categories provided earlier, it is easy to make connections between funding sources and business models: Having an endowment and core support from foundations could be linked with academic think tanks; income drawn from the sale of products or services, as well as support from corporations, and project-driven work from international donors and agencies could be linked with contract think tanks; and individuals' support and funding from political foundations can be linked with advocacy centres.

Unfortunately, these categories cannot be taken to be clear-cut or exclusive. Some think tanks are able or de facto forced to work in more than one or all four ways. Attempting to explain how they arrived at their own mix of business models offers an interesting opportunity to unearth the effect that external political, economic, and social forces have on think tanks. Srivastava, for example, described how the retreat of the public sector in South Asia and the subsequent increase in donor funding changed the balance from academic and research-driven to contract and policy-driven work (and in some cases promoted the formation of new think tanks entirely focused on consultancy or NGO-style work) (Srivastava, 2012).

Similarly, most think tanks appear to manage to balance all models rather well — even if it has a cost on their capacity to adopt and develop.

A project developed and implemented by CIPPEC shows how: Agenda Presidencial[1] was an opportunity to mobilise the think tank's own funds, drawing from the knowledge gained over a decade of research undertaken mostly via short- to medium-term contracts, to develop an initiative to inform the policy debate during the 2011 Argentine presidential elections. This involved a significant investment in communication and advocacy activities and close engagement with political parties and leaders. Furthermore, many of CIPPEC's policy proposals were supported by evidence from pilots that they had implemented with several local governments.

But as the case of CIPPEC shows, these models did not all come into play simultaneously. It provides an excellent example of how think tanks can combine all four modes of work depending on the needs of the organisation and the opportunities it faces. Therefore, labelling an organisation's business model must take this inter-temporality into account.

The SMERU Research Institute in Indonesia and ZIPAR in Zambia also enjoy a combination of core and contract funding. In the case of SMERU, core funding has commonly covered its central costs while contract funding sustained its research activities. ZIPAR, on the other hand, like many other African think tanks associated with the ministries of finance, benefits from both core and contract funding for its research — core funding directed at more academic outputs and contract funding towards more practical analysis and the development of specific policy advice or solutions. The same was true for all the think tanks supported by the Think Tank Initiative. All received, at least from the initiative, long-term institutional funding, as well as project funding from other sources.

The picture that emerges, even when long-term funding is available, is that the relationship between funders and think tanks is highly conditional. Despite efforts such as the Think Tank Initiative and the core funding offered by organisations such as the African Capacity Building Foundation (ACBF), the contract culture remains strong and dominates most of the organisations' work.[2]

---

[1] Agenda Presidencial is part of a group of Latin American think tanks' initiatives to inform electoral processes. They often publish advice aimed at political parties and their candidates, broad engagement with the media, and, sometimes, organising or promoting presidential debates. See: https://onthinktanks.org/series/think-tanks-and-elections/.

[2] This should in no way be interpreted as an argument for more "free money" for all think tanks.

But there are always exceptions. The African Centre for Economic Transformation (ACET), riding on the reputation of its founder and first director, K.Y. Amoako, received almost all its funding in the form of unaccountable grants to undertake independent research and deliver advice and support to African governments. Slowly, though, its funders are beginning to demand more accountability, which some think tankers interpret as nothing more than another way of demanding logframes and contracts for the provision of specific services.

Drawing a boundary around the think tank label

Carefully describing the balance between independent research (whether partisan or not), contract work, advocacy, and implementation and think tanks' relationship with their funding sources and the type of funding received is, in my view, a more accurate method for classifying different organisations in a way that accommodates for contextual characteristics. It also leads to interesting questions about where the boundaries of the label lie.

For example, inspired by this discussion, it is possible to draw a boundary between think tanks and research consultancies. Organisations that receive funds to provide a service to the same funder are different from those that receive funds to provide a service directed to a third party: The former is a research consultancy while the latter could be a think tank. For example, the first case could be illustrated by a contract from a development agency to review and offer advice on its own aid effectiveness strategy and the second by a contract from a development agency to review another actor's strategy and to seek to influence it. In the first case, the funding agency is clearly a client; in the second, the funding agency could still be called a funder or donor depending on the type of contract involved.

Another boundary could be established between organisations that advocate for or implement projects based on ideas developed by others and those that base their advocacy and implementation on their own research and analysis. Even advocacy think tanks ought to be in control of the ideas that underpin their work, whether because they were developed by the organisations themselves, because they played a role in their development, or because they are actively monitoring and learning from their implementation. If they are not, then they may not deserve the label.

A third boundary could refer to the effort awarded to non-research-related work (whether that is through independent or contractual ways of working). A clear emphasis on advocacy and implementation may suggest that the organisations could be better described as advocacy non-governmental organisations (driven by sometimes explicitly stated values or interests rather than research and analysis) or non-for-profit management consultancies (which are still based on interests and values, but may be more latent or tacit).

At the same time, we could question whether, without an explicit organisation-level commitment to advocacy or communications, an organisation should be called a think tank. It may be more appropriate to recognise it as an academic research centre if the focus is on independent research, or a research consultancy for contract work. Both fulfil important roles in any policy community and so should not be dismissed.

Human Resources for alternative ends

The balance of work or business model classification also opens up the analysis of think tanks' human resources (Braml, 2004; Kwaku Ohemeng, 2005) as a way of further describing their unique characteristics. Weaver (1989) and Struyk (2006) provide detailed studies on staffing choices, including the decision to employ in-house staff or external consultants or associates. Staffing is an issue of great concern and is frequently reported as a key challenge to growth by think tanks in developing countries: both in relation to questions regarding the right balance of skills and how to attract and retain the best-qualified candidates.

Staffing therefore often features in conversations with think tank directors on whether an organisation should be called a think tank or not. For example, since the COVID-19 pandemic, many think tanks have adopted more flexible working practices. Some now argue that an organisation should still be considered a think tank even if it does not directly employ researchers but instead subcontracts them, draws on the work of others, repackages this research to address key policy issues, and focuses on communicating it. Only a few years ago, this opinion would have been shot down across the think tank community.

Conversely, others would now argue that if an organisation relies on the communication capacities of other organisations, as it employs no communicators itself, the think tank label may not be appropriate.

An important distinction that think tanks like to make is between them and networks. For many of the think tank directors in developing countries and their funders that I have spoken to, organisations whose researchers are not full-time employees of the centres or whose staff is not mainly affiliated with it should be better described as networks or coalitions. This is an interesting distinction because many think tanks, particularly the smaller and more politically engaged ones, tend to rely on non-permanent staff and draw their research and researchers from a range of organisations including other think tanks and universities.

The capacity to bring in staff from outside the think tanks themselves is important in understanding their comparative advantage and value. Think tanks are often (and I describe this further in the following) considered to be intermediaries or boundary workers between the spaces of academic research and policymaking. This, according to Robert Hoppe (2010), would imply that think tanks ought to employ staff with experience and competency in both research and policy — which may explain why some prefer to draw from a network of researchers or collaborators normally employed elsewhere in policy and research organisations. Could this be a staffing condition for organisations wishing to be labelled as such?

Intellectual autonomy and affiliation

Another aspect of the strategy of a think tank is the manner in which the research agenda is set and developed. Braml, comparing think tanks in the United States and Germany, found that researchers have more freedom to choose their research agenda in academic or independent think tanks than in advocacy ones, where senior managers tend to be in charge (2004).

This matches with other experiences from think tanks in other contexts. More academically oriented contract think tanks, such as GRADE in Peru, where researchers have been employed precisely for their academic qualifications, afford employees relatively more liberty to outline their own research agendas and develop them. At the Centre for Trade Policy and Development (CTPD), in Zambia, a think tank that was more focused on contract and advocacy work, or at CIPPEC in Argentina, individual researchers and analysts have to negotiate their agendas with the centre and are, in fact, strongly reliant on their directors for guidance.

The type of funding can also have an effect on the autonomy to define and develop research agendas at the organisational level. Since CIPPEC was able to mobilise funds from individual and corporate donations, it had the power to set the agenda with its Agenda Presidencial project in 2011. This initiative involved an organisation-wide effort to develop a comprehensive government proposal for the presidential Argentine 2011 presidential candidates.

This is not always possible. In Indonesia, according to its director, the research market in which SMERU operates (certainly until 2013 before the launch of AusAID's Knowledge Sector Initiative) was increasingly forcing the organisation to focus on responding to demands from donors or the government, and this sometimes limited its own intellectual freedom. SMERU has responded to this by developing a strategy to build up sufficient reserves to carve up a space for its own initiatives.

There are other factors that constrain intellectual freedom. The source or origin of an organisation's main messages or arguments can explain a great deal. Stephen Yeo, former director of the Centre for Economic Policy Research (CEPR), suggested that the arguments of think tanks might emerge from the following (Mendizabal, 2011b):

- Ideology, values, or interests.
- Applied, empirical, or synthesis research.
- Theoretical or academic research.

I will come back to this point when I address the relational approach to defining think tanks but it is not difficult to start looking for examples for each: Ideologically (often on the "left," "right," or "centre" of the political spectrum) aligned centres, or legacy centres (affiliated with an individual with political aspirations), smaller and short-term analytical centres, and university-affiliated or hosted organisations are present in almost every context.

Josef Braml has introduced the idea that autonomy, rather than independence, is a more relevant concept for think tanks. After all, a public think tank with the mandate and funding certainty to be innovative is more likely to be autonomous than one funded by individuals and foundations from the liberal or conservative ends of the ideological spectrum and charged with advancing their beliefs, despite how "free" the funds are said to be. The concept of autonomy also opens the possibility of including organisations, which, although dependent on others for certain services (for example, research centres within a university), may still be able to maintain their own intellectual agency, and by definition, an autonomous think tank would be free to affiliate itself with any particular ideological position.

My engagement with think tanks in China and Saudi Arabia, for instance, adds weight to this argument. Think tanks that are clearly part of the regime and play pre-defined functions can still be intellectually autonomous.

The question of ideological affiliation raises a new and more interesting challenge: How different are the formal and informal affiliations with political or interest groups — and should we treat them so? Few think tanks are formally affiliated with a particular party or candidate, as this would compromise their legal status as non-profit organisations. There are, however, many examples of informal affiliations, in which the link is more ideological, personal, and historical (Mendizabal and Sample, 2009; Belletini, 2007; Garce and Una, 2007). Examples include members of a political party setting up a think tank with the primary intention of informing their party's strategy, former leaders establishing think tanks to defend their legacies, and cases with close professional connections between think tank researchers and party members and leaders.

To add to this complexity, think tanks' affiliations can be considered at the level of individuals. For example, CEPR in Europe, the Instituto de Estudios Peruanos (IEP) in Peru, and the African Policy Research Institute (APRI) in Berlin have models more akin to what Stephen Yeo has referred to as think nets: Many of their researchers are all affiliated with other research organisations, most commonly universities.

To incorporate affiliation, then, we should add Braml's *party think tank* (Weaver, 1989) category to the others identified by Correa and Botto and discussed earlier. Driven by their ideological affiliation, Braml suggests these four further types of think tanks:

1. Ideologically non-identifiable academic and contract think tanks.
2. Ideologically identifiable advocacy and party think tanks.
3. Ideologically non-identifiable advocacy and party think tanks.
4. Ideologically identifiable academic and contract think tanks.

These categories incorporate both party think tanks, those formally of a party such as the Central Research Department of the Conservative Party in the United Kingdom, and those friendly or aligned to a party's ideological position, such as Res Publica or Policy Exchange, closely linked through personal and policy ties to the Conservatives (Snowdon, 2010). Many of the think tanks that emerged in Chile during the 1970s and 1980s were also informally affiliated with the parties that made up the

Concertación government in the 1990s and had a clearly identifiable ideology (Puryear, 1994). In Zambia, PMRC belongs to the latter group, closely linked yet independent.

Within the party structure (in a single- or multi-party context), these can be as intellectually independent as any non-partisan think tank, and in fact would often compete with them to influence the political agenda. But should the fact that they pay more or exclusive attention to a particular political actor and not to the public at large affect whether they are described as think tanks? Or are they better labelled as parties' research centres? Or is it simply a matter of adding a "Party" or "Political" prefix to the label to differentiate them from the rest?

In Peru, certain think tanks are more closely affiliated, albeit informally, to parties and governments perceived as being from the left or the right of the political spectrum. For example, several researchers from CIUP worked as ministers for Alan García's market-friendly government (2006–2011) and in many cases led the development and implementation of policies that are perceived to be, by the Peruvian public, on the right. Ollanta Humala's government (2011–2016), on the other hand, has drawn ideas and experts from other think tanks such as GRADE or IEP as well as research centres from the Universidad Católica whose affiliations are sometimes perceived to be left-leaning (although changes in the president's ideological positioning may see some changes in his government's links to think tanks). While this does not mean that the centres themselves are conservative or liberal (or on the right or left), it does show that the specific roles played by some researchers can create a strong impression of a common purpose or affiliation for the organisation.

In Africa, it seems harder to assess the political ideological affiliation of think tanks. Emma Broadbent (2012), drawing on a series of studies on the political economy on research uptake in Africa, suggests this is because the traditional "class" divides which we associate with allegiances to political ideology are not applicable (despite a growing middle class), or because Africans researchers are weary of the term "ideology" having been brought up (like much of Latin America) on liberation theory, Marx, and neo-Marxism.

During a visit to Lusaka in 2011, I asked a few think tanks and civil society organisations about their ideological positions. I asked whether there were any organisations with which they would not, based on ideological differences, work. Their answer was a resounding no: There were no organisations they would not be willing to work with due to ideological differences. Even the appearance of a more politically savvy think tank in 2011, PMRC, created a surprising state of confusion among donors and NGOs. While labelled as an ideological think tank by some, none were able to describe their ideology.

In fact, most considered that the only actor opposing their recommendations was the government — an opposition that many times had more to do with its own incapacity or unwillingness to respond rather than a more fundamental ideological disagreement. As a consequence, market-friendly policies such as the removal of subsidies are advocated alongside large public investments or the removal of user fees in public services. This lack of ideological cohesiveness is all the more interesting when one considers that many such organisations work as both advocacy and contract think tanks. This implies that someone else is setting the agenda — in this case, international development funders.

Are intellectual dependence (directly or indirectly) and formal affiliation useful for defining the boundary of the think tank label? Braml, Belletini, Garcé, and others would disagree. They would even argue that think tanks could be part of political parties and other organisations. By that logic, then, government research centres such as those commonly found in Southeast and East Asia should be considered think tanks, too.

Influencing approaches

Another characteristic of think tanks' strategies is the type of influencing approaches followed. It is difficult to write about think tanks without considering their influencing function, so I will not attempt to cover all authors here. Abelson offers a good enough list (2009). He differentiates between public and private influence, which is, in itself, an interesting concept and one that is discussed further herein.

Public influence can be pursued (p. 78) through the following:

- Public forums and conferences.
- Encouraging scholars to give public lectures.
- Testifying before committees.
- Publishing a wide range of outputs.

- Online communications.
- Targeting the public during annual fundraising campaigns.
- Enhancing their media exposure.

Private influence can be pursued (p. 82) through the following:

- Accepting positions in government.
- Serving on policy task forces, transition team during elections, and presidential advisory boards.
- Maintaining liaison officers with the House of Representatives and Congress.
- Inviting select policymakers to participate in conferences.
- Allowing bureaucrats to work at the think tanks for a limited time.
- Offering former policymakers positions in the think tanks.
- Preparing studies and policy briefs for policymakers.

The attention provided by think tanks to private influence (closed and free from political meddling, the media, or the participation of non-experts or intermediaries) is stronger in the literature and experience from the Global South. The roles of patrons in China (Li, 2009) and South Asia (Srivastava, 2012), for example, and the importance of technocratic networks or informal policy communities in Peru (Tanaka *et al.*, 2009) and China (Li, 2009), confirm this private focus.

Moreover, the pressure on think tanks funded by the international development sector to demonstrate their impact to their funders has the (unintended) effect of reducing incentives to open up the debate or make use of public intermediaries, since tracking influence would then be more difficult (i.e., attribution problem).

The influencing approach could also take into account think tanks' time horizons as suggested by Ricci (1993) (Weidenbaum, 2009):

- Long-term mobilisation would include writing academic journal articles, books, and disseminating scholarly studies to thoughtful readers. This could also include the formation of coalitions of practice or teaching assignments at universities, thus helping to form future generations of policymakers.
- Short-term mobilisation would rely on intermediary activities such as participating in TV debates, writing op-eds, sponsoring a variety of events, or testifying before congress or parliament.

Publics or audiences

Finally, the choice of public (for instance, policymakers, the media, other researchers, or the public in general) can help differentiate think tanks further. Again, not wanting to review all possibilities at this point, let me turn to an interesting classification drawn from the study of think tanks in China. Looking at data from 301 think tanks in 2004, Zhu Zufeng suggests three levels of think tanks' influence (Zufeng, 2009) at which different types of public are targeted:

- Decision-makers, in other words, the individuals in charge of making the decisions, big or small, that make up the policymaking process.
- The centre, or the social elites and mass media that directly surround decision-makers and provide the immediate context within which decisions are made.
- The periphery, the general public, which is important, but has fewer opportunities to influence decision-making.

This is a useful classification because it can help explain why think tanks that focus on decision-makers themselves are less likely to pay attention to mass communications and more likely to seek to hire more senior and well-connected researchers than those interested in affected public opinion. These would be more likely to employ more communicators and researchers with strong connections with civil society organisations and other intermediaries such as the media or business and professional associations. The choice of public then can help explain the influencing approaches, the staffing strategy, the business model, and funding preferences, among others.

Some think tank directors have confided that they consider that organisations that do not seek to inform the public at large (the periphery in Zufeng's classification) more as consultancies than think tanks. This sentiment is certainly echoed by think tanks more closely associated with NGOs and advocacy, whose backgrounds are in grassroots or service delivery initiatives. More academic think tanks, on the other hand, tend to be more closely associated with influencing small groups of political, economic, or social elites, some even focusing entirely on technocrats.

In general, these characteristics of think tanks strategy offer more insights into how think tanks make use of their resources, they help to tell them apart from each other, and present us with interesting questions about their nature. They do not, however, go a long way to explain why they may be important and how they contribute to their societies. The relational and functional descriptions below address these two questions.

Relational Definitions

Following Tom Medvetz's emphasis on self-identification, another way to address the problem of definition is to do away with all these questions about rigid boundaries and instead let organisations define themselves and see which fall under the think tank label and which don't (2008). Not only would we find a greater diversity among them but we would also identify interesting cases at the edges, where the think tank space overlaps with other and where the organisations themselves find it difficult to choose a label.

Medvetz positions the space that think tanks inhabit at the intersection of the fields of politics, knowledge production, economics, and the media. A think tank, according to him, defines itself not only as one but also as "not other organisations." In 2010, ODI's director, Alison Evans, described it as a *think tank with a twist* to emphasise and explain the organisation's unique and often contradictory nature. These contradictions stemmed from the fact that, at the time, ODI was increasingly competing and collaborating with the organisations that it was, at the same time, trying to differentiate itself from: consultancy firms, NGOs, university research departments, and the media. This relationship reflected converging priorities and demands from international development research funders that imposed new functions on ODI and other international development think tanks.[3]

Finding the space that think tanks inhabit can be a useful exercise to explain what they are and what they do. Based on a conversation with Stephen Yeo, Figure 1 attempts to place think tanks in a broader community of organisations that behave in similar ways. The shaded space is

---

[3] Typically, bilateral donors, multilateral organisations, and foundations increasingly demand large, multi-country, multi-year policy research consortia with strong emphasis on policy influence, capacity development, and implementation. Think tanks may find the space in these consortia reduced: Academic think tanks and university research centres are better suited for the research component, NGOs and advocacy organisations are better candidates for the policy-influencing component, and consultancies are better prepared to manage complex consortia and the capacity-building and implementation components.

Mode of work \ Message based on	Ideology, values, or interests	Applied, empirical, or synthesis research	Theoretical or academic research
Independent research	Public intellectuals/Opinion leaders Party think tanks		Academic research centres in Universities
Consultancy/contract	The media, interest groups, NGOs, lobbies	Applied research centres in universities and academic think tanks Advisory bodies within the civil service Consultancies including contract think tanks	
Influence/advocacy	Advocacy think tanks Political Parties		Chief scientific advisors, Public intellectuals/Opinion leaders

Figure 1. Think tanks in context by mode of working and source of arguments

shared by a number of organisations that may or may not label themselves as think tanks but certainly compete for the label. If this is the case, then the study of think tanks ought to focus on both think tanks as well as on the relationship between think tanks and other players.

Because they have to interact with these other organisations, think tanks construct themselves alongside complex and unpredictable paths. The choice of business model, for example, may have more to do with the political and funding contexts in which they operate than their desired choice of mission or original values.

In pursuing a research niche or influencing approach, they simultaneously seek to differentiate themselves while adopting practices from others in the policy space in which they work. It is not uncommon, therefore, to find that although some think tanks present an image of applied research centres, they also adopt certain practices from academia, such as appointing "fellows," publishing academic journals, and establishing highly cumbersome quality control processes that in fact limit their own ability to function as think tanks. Equally, they adopt management and marketing or communication approaches probably more appropriate for private sector consultancies or the media than for research centres.

Hoppe's boundary workers concept provides further support for this proposition. He places think tanks at the centre of several communities or spaces, suggesting that they must simultaneously belong to them while maintaining a separate identity, one that is different from the organisations that only belong to those other spaces (Hoppe, 2010). The keyword is *simultaneously*. This makes the definition of think tanks particularly difficult. If we follow the arguments presented by Medvetz and Hoppe, we must recognise that think tanks should exhibit characteristics of other types of organisations. Therefore, the idea that we could trace neat boundary lines around them is quite simply unrealistic.

Functional Descriptions

While it may be sometimes difficult to point at an organisation and say, with absolute certainty, whether it is a think tank or not, it is certainly easier to say whether it behaves as one. Think tanks can also be described by the roles or functions they fulfil in relation to their context, to the politics in which they have been set up and have developed.

A functional approach makes it possible for organisations with entirely different funding structures, business models, strategies, affiliations, audiences, and relations to be labelled as think tanks if they carry out the functions expected of one in each context. Similarly, organisations that may look like a think tank on paper could be easily excluded if they fulfilled none of the expected functions. In my view, this makes it possible to talk about think tanks in China, the United States, Peru, and Indonesia without stopping to argue about their obvious organisational and contextual differences.

The functional approach also accounts for the effects that the political, economic, and social contexts have on think tanks' origins and development, and which may not have allowed them to develop the idealised characteristics they may have wished to.

This is not a new idea. It was put forward for the study of think tanks and political parties in Latin America (Mendizabal and Sample, 2009) drawing from research on networks (Mendizabal, 2006b) and Orazio Belletini's work on think tanks in Ecuador (Belletini, 2007). At the time, the functions suggested were rather closely related to a range of strategic influencing approaches. Since then, I have expanded them to incorporate other functions described in the literature about and related to think tanks and practices from think tanks in developing countries. Others have also used the approach in an attempt to introduce the idea of think tanks to a wider audience.

In April 2010, Kanti Pajpai wrote in the *Times of India* an article about the roles that think tanks could play in "rebooting" India's democracy (Bajpai, 2010). He suggested a number of functions that think tanks could play:

- **Policymaking function:** "The first is to help create policy where there is none. A think tank may direct government and public attention to an emerging or a neglected area of social life, which requires policy intervention."
- **Paradigm shifts:** "A second function is to fundamentally change the direction or nature of existing policy by means of a paradigm shift. It can do so by showing that the original conditions that brought forth a policy intervention have changed or that existing policy is ineffective."
- **Adjusting direction:** "Third, think tanks can help modify existing policy for the same kinds of reasons changed conditions and lack of effectiveness."
- **Monitoring and auditing:** "A fourth role is to monitor existing policy to see if it is implemented properly and to bring success and failure to the attention of the authorities and the public."
- **Public education:** "Then, think tanks have an information role in respect of the larger public. They may simply disseminate to ordinary citizens, without critical commentary, what the government is doing in various areas of social policy and educate the man on the street the nature of various programmes."
- **Prospective:** "Finally, think tanks can incubate ideas for the future. This is a vital role, one that focuses not on immediate policy concerns but rather has a more distant horizon. It is also perhaps a more theoretical function in the sense that the think tank in this role is concerned with constructing a whole new vocabulary and set of conceptions about various areas of social life with perhaps no immediate relevance or application."

Taking these into account, I would outline the following functions or roles for think tanks:

(1) They help to identify problems to be solved. In Spanish, it is said that one needs to *problematizar* (explain the problem) an issue before dealing with it. Think tanks help to identify new problems by explaining their causes and effects on society, and drawing attention (and resources) to them by making them the object of research and policy (Belletini, 2007; Clifton, 2010). This is a role that think tanks can fulfil particularly well given their connections with actors in multiple spaces: They can track new research and ideas emerging from political debate, scientific research, corporate affairs, and so on. Their participation in international networks and initiatives can also help to consider problems that may have been identified elsewhere but may be quite relevant to their own contexts. Their links to other civil society organisations can also keep them closely aware of the challenges faced by ordinary people.

A certain degree of intellectual autonomy is necessary for this function. For instance, the Jesuit Centre for Theological Reflection (JCTR) carried out monthly surveys to calculate the value of a basic needs basket in twelve cities and towns in Zambia. This exercise provided an opportunity to identify areas or issues that require attention from researchers, civil society, and policymakers. Most importantly, it was rather successful at, through a simple monetary figure that everyone can understand and relate to (the cost of food, transport, housing, etc.), drawing the attention of the general public to basic common "needs."

(2) Think tanks create, nurture, broker, and promote policy ideas. In other words, think tanks are incubators and repositories of valuable resources for policymakers and other interested policy players (Abelson, 2006, 2009; Clifton, 2010; Mendizabal and Sample, 2009; Tanner, 2002; and others). It is not just a matter of conducting research and communicating it to legislators or ministers but also identifying useful new insights from academia and promoting the most relevant ideas (often adapted to address current problems and concerns).

(3) They develop the capacities of researchers, public intellectuals, and policymakers: Ideas come from people, such as experienced policymakers taking some time off to reflect on what they may have learned, or young policy entrepreneurs eager to join the policymaking ranks.

Think tanks can provide the institutional support they need to develop their skills and arguments.

This is a point that very clearly emerges in several conversations with both young and experienced think tank researchers. In many developing countries, where the university system is weak, think tanks offer the only opportunity for young graduates to practice and learn policy research and analysis methods. They also present them with opportunities to engage with policymaking, as many researchers are close to or also acting as policymakers or politicians.

Think tanks can be seen as a stepping-stone in the professional development of future policy analysts, political advisors, and other invaluable participants in effective policymaking systems (Braml, 2004).

(4) They create, mediate, and open new spaces. Jeffrey Puryear's account of think tanks' roles in Chile is particularly relevant (Puryear, 1994). Despite all the evidence that they played an instrumental role in the definition and implementation of the opposition's strategy when many think tank staffers joined the new government in the 1990s, Puryear argued that the convening function of think tanks was their most important contribution. He calls it the think tanks' psychological contribution. Chilean think tanks' events and meetings during the 1980s, and in particular the manner in which they were conducted, helped to create new relations between the various opposition parties and trained their leaders on democratic practices. At a more fundamental level, think tanks became spaces where ideas and their researchers were safe from the repression of the military regime (Cociña and Toro, 2009).

(5) Think tanks have also been described as vehicles or windows into and out of the policymaking *black box*. Tanner (2002) and others writing about Chinese think tanks have stressed their role in providing a view into the Chinese political space, as well as allowing Chinese officials a better view of the rest of the world. The rather closed nature of the Chinese State not only meant that it was difficult for Western policymakers to find out what was going on in China but Chinese policymakers found it equally difficult to learn about the rest of the world. Hence, state-sanctioned think tanks and experts (and increasingly private ones) have been used as windows to the outside world and have fostered the development of new relationships between Chinese and foreign policymakers, scholars, and interest groups. As suggested

by Nachiappan *et al.* (2010) and Cheng Li (2009), this is not a matter of chance, but of planning, in this case, by the state (or an elite within the state). In fact, think tank researchers in China have been used to carry specific messages to foreign officials and to influence foreign scholars with whom they have close ties so that they would, in turn, influence policies in favour of Chinese interests (Shambaugh, 2002).

The window is probably a limited metaphor as think tanks do a lot more than just letting the light through. Robert Hoppe's concept of boundary workers is more accurate (2010): Think tanks, through their often-porous relations, actively and simultaneously participate in multiple spaces, thus allowing their members to engage with people, ideas, and processes that they would not have otherwise been able to. This connecting quality led Braml to describe them as an *organisation homo mediaticus* (2004, pp. 25–26) that contributes to a pluralistic and democratic society.

(6) They can also serve as channels of resources and support to other political players and interest groups (Abelson, 2009; Weidenbaum, 2009; Belletini, 2007). The close formal and informal links that exist between think tanks and political parties, as well as other interest groups, and the nature of the sector, in which parties can be seen in essence as consumers of research and advice, mean that their work is often supportive of their political ambitions. Some, particularly internal think tanks or think tanks directly supported by political and ideological foundations, can be more clearly described as channelling resources to political players, for instance, transforming funds into policy options and ideas for political parties to use in their campaigns or even in government or opposition duties.

They help legitimise policies and personal, organisational, or national ideologies, interests, or *brands*. The study of think tanks in East and Southeast Asia (Nachiappan *et al.*, 2010) illustrates the manner in which think tanks are used as legitimators of policy decisions made by politicians or other interested parties. My research on DFID's use of evidence showed that evidence is largely used to make sense of political demands and commitments rather than as an initiator of policy (Mendizabal and Jones, 2010). But the literature on think tanks also offers an insight into the formation of think tanks as the tools or vehicles of political hopefuls or leaders in search of a technocratic credential, but in many cases these promoters are development agencies themselves. They see think tanks as vehicles to promote development and, in some cases, as outputs of their aid interventions. Think tanks in Malaysia are often linked to political hopefuls or established political leaders.

In the international development sector, the practice of using local think tanks is also partly fuelled by the search for legitimacy. Campaigning NGOs also employ think tanks to develop technical reports backing up their policy demands.

(7) At the same time, think tanks can fulfil invaluable social and political monitoring and auditing functions. Hugh Gusterson (2009) presented a strong argument for the recognition of think tanks as an auditor of public or private policy and behaviour. In fact, many think tanks, whether operating as independent researchers or contracted by others, undertake policy evaluations, develop indices to monitor the progress of policies or problems, and play an active role in raising the alarm on key issues of public interest that may be being overlooked by those responsible. The monitoring role is particularly common among "opposition" think tanks and international development, and developing country think tanks whose funding is often associated with budget monitoring, good governance, and accountability interventions.

In Peru, for example, conversations about think tanks often lead to the question of whether the many *observatorios* that have been set up in the country fit the label. These are monitoring programmes (commonly set up as networks or sometimes as standalone organisations) that focus on a policy issue or sector (e.g., health or education), public spending more generally, human rights and corruption, or, more recently, local governments and cities. Strictly speaking, these are not think tanks, but they fulfil functions that some think tanks may choose.

Over the last decade, as governments have become more authoritarian and the civic space is closing, think tanks have opted to adopt a watchdog function. Liliana Alvarado from Ethos in Mexico explains this is a role that think tanks need to adopt as the context changes (Alvarado, 2023).

This role however is difficult to marry with others, as it would demand that the think tanks and their researchers remain entirely independent from competing interests.

(8) Often criticised as elitists, think tanks can also carry out a very important elite and public education function. This is a role that has existed for more than a millennium: Political advice in the Western world began when famous philosophers tutored young princes, Aristotle

tutored Alexander, Seneca taught Nero, and later Thomas Hobbes tutored Charles II (Smith, 1991). The modern think tank fulfils a similar role: The Brookings Institution has a number of initiatives to support incoming congress members; CIUP in Peru, CIEPLAN in Chile, and CIPPEC in Argentina have worked with political candidates to inform and prepare them on the key demands for government and policymaking. Think tanks across the world participate in one way or another in advisory panels, committees, or councils for presidents, ministers, and parliaments.

The aspect of public education however is one that seems to receive less attention. The emphasis that international research funders have placed on demonstrating influence over the last few years means that think tanks are under pressure to move away from engaging in the indirect, slow, and long-term function of educating the public on matters of public interest. However, popularising complex and technical ideas and arguments to encourage the general public to participate in policy debates is a central role for think tanks. Ricci argues that think tanks' core purpose should be to contribute to what he calls the Great Conversation, rather than simply focus on private dealings with the political and economic elites (1993).

The functional approach demands that when we consider who deserves the label of a think tank, we first attempt to understand the purpose of the organisation. Questions about funding, business models, staffing, and others may then make more sense. Should an organisation be labelled as a think tank if its design is focused on income generation and driven by projects? It might be best to describe it as a consultancy that publishes and communicates research (McKinsey, for example). Would an organisation whose staffing choices respond to academic and pedagogical objectives above all else be considered as a think tank, or is it better to label it as an academic research centre? We could ask the same about organisations that share the purposes of the media, political parties, or policymaking bodies.

Are We Wiser?

What, then, is the right way of defining and describing think tanks? We have discussed the normative definition. This is broadly based on the legal code and draws from the Anglo-American think tank tradition. According to it, a think tank is an independent non-for-profit organisation, free from financial or ideological constraints.

This definition, however, is limiting for the study of think tanks and tends to impose ideals that may not be feasible for most developing country think tanks. It is also limited in that it is not specific to think tanks but to a wider set of civil society organisations. And as such, it does not recognise those that are part of the government, parties, the corporate sector, or even international bodies.

Organisational categories (related to size, for instance) are equally limiting but better at providing us with useful information about the nature of specific think tanks. This often depends more on their context than on their own agency. So, comparing think tanks across regions or countries based on these categories can be misleading. Even within a country, think tanks targeting different policy spaces or working on different policy issues may exhibit important organisational differences. Organisational characteristics move us further from the limiting normative definition, but do not account for the reasons for these differences.

Focusing on the agency of think tanks offers an attractive alternative. It unpacks a much broader set of characteristics and opens several discussions about the boundaries of the concept of think tanks. While funding models, business models, staffing strategies, affiliations, influencing approaches, and the public or audiences they seek to influence are certainly affected by external forces, they still constitute choices made by the think tanks themselves. These choices can help to set them apart from each other more accurately than normative definitions or measurements of size or might. They can also help us to tell them apart from other types of organisations with which they may share some organisational and functional similarities.

If anything, these characterisations illustrate how hard it is to define what is and what is not a think tank. The exercise, in the end, appears to be self-labelling and relational at its core. Of course, self-labelling also describes agency: An organisation chooses to label itself as a think tank because it suits its own interests. However, this too can be mainly externally driven. The prospect of accessing the funds offered by the Think Tank Initiative, for example, drove many research centres to rebrand themselves as think tanks, even if they had reservations and concerns about what that would imply. In other cases, it is the adoption of the label by donors and other international development agents that has permeated the way organisations describe themselves.

This self-labelling or imposed labelling is entirely relational. In places where there are no easily identifiable think tanks, organisations that share some of their characteristics or functions have been quickly labelled as such, even if they would be unrecognisable as such anywhere else. But the label carries a certain degree of responsibility, and it triggers an effort to engineer differences with other policy actors and possible competitors. This, as we saw in the discussion on agency, poses serious questions about where the boundary lines ought to be drawn.

Finally, this brings us back to think tanks' functions: What are they for? Different forms of organisations that have made different choices in relation to internal and external forces may still share key roles or functions that can be attributed to think tanks. While this is still my preferred approach to describing think tanks, the discussion offered by the other approaches introduces new challenges.

We must still answer the basic question regarding the main purpose of a think tank. What is its core or ultimate *telos*? Should an organisation that only provides an auditing function be called a think tank? What about the role of training the next generations of policymakers? Or the function of creating spaces for debate? They are all important, but it does not seem right that these, on their own, may be sufficient to qualify as such.

Universities are much better than think tanks at developing new generations of researchers and policymakers; to train and prepare young people is their core purpose. The media is much better at creating spaces of engagement and debate to inform the general public, as well as taking on the role of auditors, too. Many other public bodies and civil society organisations can fulfil similar roles, as well. These are intrinsic to their own claims for their legitimate participation in public life. Therefore, a certain normative element for think tank functions is needed, something akin to a mission statement that outlines their most fundamental purpose.

In my view then it is not enough that they may fulfil one or more of the functions outlined herein. In fact, these can change and gain or lose importance depending on the different contexts in which think tanks are found, and, as we saw in the case of business models, these can also change over time. Think tanks ought to, at least, want to affect the decisions of public and private actors on issues of public interest. This is something that all think tank directors I have talked to agree with. How to do it, however, does not enjoy the same level of agreement. I would suggest that to affect the decisions of public and private actors and to be labelled as think tanks, they should significantly draw from their own research and analysis (but also from other sources) to identify and develop new arguments (including the identification of problems and the development of solutions) and seek support for these to be acted upon.

The functions listed herein describe the different approaches that think tanks can follow to fulfil this mission. In doing so, I would again argue that they should not allow their formal or informal affiliations, as well as their business or funding models, to entirely constrain their intellectual, ideological, and operational autonomy. If they do so, then maybe another label may be more appropriate.

A final key characteristic of a think tank that I would like to propose, for the purpose of this study at least, is that they must label themselves as such and that this should go beyond a simple branding exercise driven by senior management or some public relations decision in search of new funding sources. Self-labelling goes a long way in deserving the label: An organisation that wants to be a think tank certainly deserves the label more than one which does not. Furthermore, in a context in which the term is fluid, what better approach than to rely on those who seek to appropriate it to inform its ever-changing boundaries?

In the meantime, though, I think I'll know one when I see one.

Section 2

Perspectives for the Study of Think Tanks

Chapter 3

From Waves to Perspectives

The literature on think tanks is heavily influenced by the idea of *waves* or *traditions* of think tanks put forward by a number of authors, most prominently Diane Stone (Belletini, 2007; Stone and Denham, 2004). The former refers to the existence of at least three waves of think tank development: from a few state-centric centres (often set up by governments themselves), to more diverse think tank communities with greater links outside the government and national borders, and finally to a situation where think tanks are in essence acting transnationally. The concept of traditions refers to the existence of regional characteristics or development waves (in Latin America, Africa, Asia, etc.) that can be helpful in the study of think tanks. Various authors have adopted these ideas, such as Orazio Bellettini, James McGaan, and even myself (for instance, with Ajoy Datta and Nicola Jones, 2010, in the assumptions behind the commissioning of several regional studies on think tanks while I worked at ODI, and throughout 15 years of writing on the On Think Tanks blog). However, the idea of a tradition does not fit nicely with what we find in reality: As we will see in the following sections, it is possible to find countless examples of stark differences between think tanks in the same regions and countries — as well as similarities between them across diverse contexts.

Moreover, the idea of development waves, particularly the suggestion that think tanks are now acting transnationally — more so today than they ever did — may very well more closely reflect the reality of think tanks in higher-income nations than the rest of the world. Or, it could also be explained by more visible support from international donors that are particularly demanding of think tanks' visibility and participation in regional or global policy spaces, as well as the changing nature of think tanks' communication channels, which now make it possible to reach global audiences — but more by the agency of their audiences than their own.

It also contradicts the historical evidence: Chinese think tanks, from their very beginning, have been oriented towards learning about the world outside China and have had to collaborate with others to do so. Chilean think tanks would not have been possible without the support from foreign funders, universities, and think tanks. The metaphors that have inspired and driven the formation of think tanks in developed countries have played central roles in developing country think tanks' national histories. In other words, this transnationality is by no means new.

Furthermore, many of the differences that we find among think tanks can be better explained by the way in which we understand and study them — rather than actual differences between them. Think tank managers and their researchers are not necessarily aware of their "type" of think tank, but they do have an idea of what they would like their organisations to look like. Through different means, they steer their organisations in the direction that best fits with their view of the world and the roles they expect their think tanks to play.

The ideas and tools we carry with us shape our responses to our environment, and this is the same for a think tank scholar. For those who believe that policymaking is the prerogative of elites, think tanks are likely to be seen and explained as members or instruments of these elites, but to those for whom policymaking is the consequence of negotiations between a plural set of actors where nobody has absolute power over all others, think tanks are probably just one more actor — barely worthy of study. As a consequence, the objectives and functions that they set out for their own think tanks are different, too. It does not come as a surprise, then, that finding a single definition of think tanks has proven to be such a challenge.

An alternative therefore is to combine these ideas with perspectives of how politics work — and the roles that different political players play, think tanks among them. Hence, I have approached this project with the explicit objective of identifying the different *perspectives* from which think tanks are being imagined and studied. This is partly influenced by several years of participation in debates on think tanks where it was obvious that our conceptual and analytical differences stemmed from our own preconceptions of the world of policymaking. I found the literature full of these, and each presented its own unique view into the world of think tanks. Abelson and others are explicit about these and have identified elitist, statist, and pluralist interpretations of policymaking and the roles of think tanks in each. Some have approached think tanks from the perspective of the organisations themselves, while others have done so through the study of the policymaking context. Others have tackled the more tangential world of public intellectuals, experts, and ideologues — and ideologies — in democratic and technocratic politics. Each perspective presents a narrative, or set of stories, from which researchers have approached the subject.

As we will see in future chapters, this approach yields some interesting results:

- Organisational, elitist, statist, and pluralist forces can play key roles even within the same country (and region) over time.
- Individual organisations can also be driven by more than one of these forces throughout their history.
- Waves of formation or development then cannot be expected to follow a particular linear trend — i.e., increased openness or *transnationality* — but reflect much more complex internal and external forces at play in the spaces that think tanks inhabit.
- Political and economic liberalisation, often assumed to be drivers of think tank formation and responsible for the so-called second wave, are in fact not necessary conditions for the emergence of think tanks.
- Even during periods of autocratic and military rule, think tanks can find fertile ground to develop.
- There are several important similarities between think tanks in extremely diverse contexts, which call into question the relevance of studying think tanks within geographic regions — or even in the imaginary "developing world."

Briefly, *elitist* and *statist* views would tend to suggest that think tanks have been engineered or planned by a hegemonic power: founded, funded, and *patronised* by international financial institutions, bilateral and multilateral donors or their staff, Northern think tanks, political hopefuls and caudillos, corporations and philanthropic foundations, and the State. For instance, in Colombia, think tanks emerged as instruments for the political groups engaged in the political and intellectual struggle to form the modern Colombian republic in the second half of the 1800s (Londoño, 2009).

Chilean think tanks emerged as a consequence of the coming together of public intellectuals, the Catholic Church, international funders, and political (in opposition) leaders (Puryear, 1994). in East and Southeast Asia, one finds a combination of corporate drivers (Japan and Korea), religious, economic, and political leaders (Malaysia and Indonesia), and the State (China and Vietnam) (Nachiappan *et al.*, 2010). In Africa, international donors and former international financial institutions' staff have been instrumental (Kimenyi and Datta, 2011).

Pluralists would opt for a more natural emergence of think tanks — as the consequence of opportunities in the marketplace of ideas: the creation of new government departments with power over certain interest groups, as well as demand for information, and the conflicts and negotiations of the democratic process and political party strategies. The idea of pluralist emergence is sometimes difficult to accept because most of the historical accounts of think tanks in the literature highlight the clear roles that certain individuals had in their foundation and development; but in fact, the vast majority of think tanks in the United States are rather small and so their appearance, although benefiting from the entrepreneurship vision of individuals, cannot be said to be limited to the actions of a small elite. The same might be inferred about the global think tank community — if there is such a thing.

Pluralist approaches necessarily emphasise the complex networks or relations that exist between different political players. This stresses the importance of self-identification at the core of think tank formation. They choose the label when, in relation to others, they come to fulfil the functions of think tanks — or when, put bluntly, it will help them attract funding (the opposite also applies: not identifying oneself as a think tank to avoid the negative connotations that the label might carry). Regardless, their place on the fringe or boundary forces them to be flexible and malleable — transforming from advisory councils into public fora into think tanks into movements.

The *organisational* perspective places a greater deal of attention upon the genesis of think tanks. However, since most of the studies of think tanks from an organisational perspective tend to be evaluations or needs assessments, the question of "why they exist" is hardly ever posed. The following sections will seek to address this, starting with think tanks themselves.

Chapter 4

Introspection: The Organisation

An obvious perspective to consider first in the study of think tanks is that of the centres themselves. Organisational approaches consider them as a unique type of organisation and therefore a worthy subject of study. Two of the best-known examples of this are Raymond Struyk's books on managing think tanks (Struyk, 2006) and James McGann's work around his "Go to Think Tank Index" (2012) — which runs up until 2020. Both centre their work on the organisations and employ different sets of criteria to describe them and, in the case of the latter, assess their influence. Unfortunately, the literature on think tanks from this point of view does not, in my view, offer the richness of other perspectives, even though it does provide useful insights into the workings of think tanks around the world. How replicable these best practices may be, however, is questionable — but certainly not irrelevant.

In any case, even in studies that take this perspective, individual organisations are seldom presented on their own. Abelson's account of the best-known think tanks in the United States (the Brookings Institution, the American Enterprise Institute, the RAND Corporation, and the Heritage Foundation) is placed within a debate on the other aforementioned perspectives. The study of the rise of foreign policy think tanks, for instance, is inseparable from the study of American foreign policy itself and of developments in the national and international political context (Raucher, 1978). Struyk's work on think tanks in Eastern Europe is only fully appreciated when their context (the effects of the Cold War, for example) is taken into account.

The same is true for research about think tanks in other contexts. Studies of Chinese think tanks place a great deal of emphasis on the evolution of Chinese politics and their effect on the immediate environment of think tanks. For instance, the opening of spaces for young economists in the 1980s to provide new insights and "intelligence" to party officials vying for power, to the crackdown on political think tanks after Tiananmen, helps explain the more recent rise of personalised think tanks patronised by well-known international figures such as Justin Li (Zufeng, 2009).

By focusing on think tanks themselves, these studies tend to isolate the changes that take place within them in the context of an evolving environment. Pluralist, statist, and elitist studies would, as we will see later, pay more attention to changes in the context and how this, in turn, may affect think tanks as well as other policy players. Raymond Struyk, for instance, focuses his work on the following key organisational issues (2006)[1]:

- Governance: including the board, salary models, staff structure, and leadership.
- Funding and finances: including financial management.
- Staffing: including staff management and productivity, and on-going staff training.
- Research and communications: including research and quality control, communication, and creating innovation.
- Business development: including commercial activities and winning work.
- Monitoring and evaluation.

Similarly, Braml has identified a number of important internal factors that, alongside the external environment (which is the true focus of his work), drive the formation and development of think tanks. Interestingly, but not surprisingly, Braml argues that the structures observed in the think tanks he studied were best described as behavioural responses to the external environment. In other words, the external environment defines the internal environment, and this internal environment is characterised by a number of decisions on key organisational issues (2004):

- The business model: described in terms of families of think tanks (ideologically identifiable and non-identifiable).

---

[1] This grouping of Struyk's categories is mine.

- The allocation of resources between research, communication, and fundraising.
- The services and product mix: long-term research, analysis, commentary, advice, training and capacity-building, etc.
- Content and packaging and the use of the think tank versus the researchers' brand.
- The identification and pursuit of distinct audiences and target groups.
- Media (visibility) and related activities.
- The distinct roles or functions assumed by the different families and types of think tanks referred to earlier.

Others, such as IDRC, are more systematic in their analytical frameworks: (Lusthaus *et al.*, 2002):

- **External environment:** including characteristics of the administrative environment, the political context, sociocultural issues, and the centre's stakeholders.
- **Organisational capacity:** including strategic leadership, structure, human resources, financial management, infrastructure, programme and process management, and inter-organisational linkages.
- **Organisational motivation:** including the organisation's history, mission, culture, and its incentives and rewards system.
- **Organisational performance:** which is explained by the earlier points and involved assessments of effectiveness, efficiency, relevance, economy, and financial viability.

Together, they help assess and understand the organisation's performance. Stephen Yeo and I used IDRC's approach for a review of the Secretariat for Institutional Support for Economic Research in Africa (SISERA) in 2003. The study found that, among other factors influencing the performance of the think tanks in the network, leadership and the organisation's culture played significant roles. But this, as suggested by Braml, only makes sense if leadership and culture are seen as a reaction or a part of the broader context. Think tanks' leadership styles and organisational culture are not independent on leadership styles and culture outside the centres.

More recent efforts to explore the organisation from a psychoanalytic theory and organisational life, such as Ajoy Datta's series on rethinking organisational development (Datta, 2021), also fail to take the rich diversity of contexts and influences on the organisation. There is an ironically normative approach to how organisations should function that does not engage with local culture, norms, or practice.

Nonetheless, this isolation of external "noise" provides valuable insights. In two reviews I conducted for ZIPAR in Zambia and the Belgrade Centre for Security Policy (BCSP), I was able to focus on the relationship between the formal governance structure of the organisation and its capacity to undertake research and policy engagement as expected. These in-depth analyses can lead to interesting findings about the relationship between think tanks and their environments.

It would not be possible to consider how the context affects think tanks unless we had a strong grasp of their nature. For example, Glaser and Saunders were able to explain the rise of think tanks in China by considering how the following changes in the context affect them and their workings (2002):

- The development of horizontal linkages that brought experts with overlapping interests together — not normal for Chinese models of vertical professional relations.
- The intensification of competition for analysis and advice, which has been transformed into more government and civilian think tanks, more funders, more researchers, etc.
- The growth in the number of better-educated and better-informed experts, including a growing number of PhDs from Western universities, skilled in foreign languages and more open-minded, which has helped think tanks find competent staff. On the other hand, foreign policy think tanks are no longer as alluring to young researchers since there are other options available to travel or to develop connections to international policy processes. It is now easier to move from one work unit to another (the marketisation of employment), and so think tanks have had to develop new incentives to attract promising candidates.
- Increased contact with foreigners, providing additional opportunities and resources for think tanks.
- The involvement of the Chinese government in the international arena, as well as the increasing complexity of Chinese political and economic life, leading to an increase in government demand for new topics of research interest, including counterterrorism and international development.
- The increasing role of university professors and institutes who are seen as more trusted and impartial — and may be so as think tanks start to compete for influence and funding.

Generally, organisational approaches tend to focus on think tanks' capacity to influence policy. There is an increasing body of literature looking at the internal and external factors that affect think tanks' influence on policy (Braun *et al.*, 2007). These studies, just as other organisation-focused ones, award think tanks a critical role in policymaking. They often take the organisation itself, or a piece of research that is deemed to have been of great value or influence, as their starting point and then attempt to describe how this influence took place and what explains it. The obvious question of whether there was any influence in the first place is not even asked. Nor is the role of other organisations or policy players considered relevant enough to be brought into the study.

As a consequence, think tanks' influencing capability is assumed to rest heavily upon their own internal organisational strengths, including good management practices, communication competency, and networks or alliances with other policy players. Their weaknesses, on the other hand, are more easily explained by unsupportive contexts — such as limited and short-term funding, uninterested or incapable policy bodies who do not demand evidence or analysis for policymaking, and media with little analytical capacity.

And so, isolated from others, the prevailing story or narrative used in the study of think tanks from this perspective is that of the bridge that must be built over the gap between think tanks and their policy audiences and objectives.

This organisation-centric perspective is predicated on (and reinforces) the recognition that think tanks are a unique kind of organisation. Their popularity can be explained by a belief that they are important in and of themselves and that they can make a difference in policy and practice. And to understand why this is the case, it is necessary to understand the history of this idea and how it has driven the formation of modern think tanks for more than a century. At the core of the organisation then is the idea that evidence can make a difference.

From Medicine to Marketing: How Key Metaphors Have Influenced the Belief in and Development of Modern Think Tanks

When I was working at ODI, I found myself, at least initially, taken over by a kind of faith in the capacity of research and evidence to solve all the world's problems — and of our role as policy entrepreneurs in promoting its use across the developing world. At the time, the year 2004, the evidence-based policy narrative was at its peak in the United Kingdom; the aid industry in Britain, driven by New Labour's Department for International Development (DFID), was one of its earlier adopters. Evidence-based policy was (and still is) a sort of mantra for the industry. ODI even set up a global network named Evidence-based Policy in Development Network as a vehicle to promote this idea.

Of course, it is difficult not to embrace it. It is a very powerful idea. And this very same idea can be credited with playing a leading role in the history of modern think tanks — one that, inspired by a never-ending train of metaphors over more than a century, has been adapted, reinterpreted, and transformed to play the same role in an ever-changing environment (Smith, 1991). What this history tells us is that the emphasis on evidence in policymaking, more recently captured by initiatives such as the Abdul Latif Jameel Poverty Action Lab (J-PAL) and the International Initiative on Impact Evaluation (3ie), is nothing new. We have been here many times before.

Since the history of think tanks in the United States has been extensively documented, it provides an excellent opportunity to introduce some of these metaphors and consider how they have affected the formation and development of these organisations. Alongside this history, I refer to think tanks in developing countries and how these metaphors continue to affect them today.

There is another reason why this particular history is important to think tanks in other contexts — especially where funding for them primarily originates in the United States or Western Europe. It explains quite a lot of the ideas behind the manner in which their funders approach and support them. Being aware of this history is, in my view, a strategic imperative for think tanks in the developing world. How else can they effectively engage and negotiate with their funders, supporters, and even critics if they do not know where they are coming from?

Society as the patient

The belief (and particularly the belief of some individuals) that science could treat the symptoms of society, just as medicine could treat patients, triggered the formation of the first modern think tanks in the

United States.[2] This belief was embodied by a number of individuals and philanthropists from the late 1880s. James Smith (1991) has described, for example, the particular role played by Richard T. Ely, who, upon his return from Germany, and having witnessed the creation of the German Welfare State, founded the American Association of Economics (1885) as a convenor of the increasing number of university-trained experts. The Association soon became a focus for presidents and officials who, equally influenced by the medical metaphor, sought advice from experts to solve specific social and economic problems.

In 1899, another think tank pioneer, John R. Commons, established the Bureau of Economic Research (which twenty years later would become the influential National Bureau of Economic Research) and then joined the National Civil Federation (1900). Together, in 1904, Ely and Commons founded the American Bureau of Industrial Research. Then, Ely set up the American Association of Labour Legislation in 1906. And the story continues. In all cases, these men were driven by the powerful idea that the scientific method could be applied to society and right all its wrongs.

This was, however, still a sort of amateur think tank community, made up of associations and networks of peers and like-minded scholars and other interested parties who volunteered their time to the community. At the time, the idea that the scientific method could be applied to society was certainly not new but had not gained the popularity it has today. These organisations, in due course, would be responsible for promoting it. They were driven just as much by public concerns and a conviction that knowledge could play a role in addressing them by the action of individuals rather than by knowledge alone.

For example, Abelson describes the appearance of the Chicago Civil Federation (1894), a predecessor of the National Civil Federation. The Federation was founded after the assassination of Mayor Carter Harrison, and growing concerns with corruption triggered the involvement of the local business and civic leaders (Abelson, 2006). The Federation established several committees to investigate various problems in local development, very much as think tanks today would seek to tackle current

---

[2] This claim is contested however, as foreign policy think tanks, arguably founded earlier that these, were motivated by other motives, namely, to keep foreign policy away from the democratic process.

problems but probably more similar to the working of advisory councils or civil society working groups.

These organisations may not have looked like the familiar think tanks of today, but several characteristics make them strong and useful references: They were independent from the state, driven by a desire to solve the ills of society and the government, and they resorted to expertise and analysis to find solutions. Still today, many think tanks in developing countries operate as de-facto associations of loosely grouped researchers and consultants, and economic associations at the national and regional level in policymaking. For example, the Consorcio de Investigación Económica y Social (CIES) in Peru and the Latin American and the Caribbean Economic Association (LACEA), as well as the Economic Association of Zambia (EAZ) and the African Economic Research Consortium (AERC), play important roles in policymaking at the national and regional levels. Like the federations formed in the United States over a century ago, they do so by bringing together researchers and practitioners to work on and around issues of public interest.

As the focus of their efforts shifted from treating the symptoms of these social maladies to their causes, these associations gave way to the formation of larger and more professional research bodies and long-term research efforts. The Carnegie Corporation and the Ford Foundation brought considerable resources to social research, but it is the Russell Sage Foundation (1907) that Smith credits with leading the new way. In an early version of participatory research, the foundation undertook surveys conceived as collaborations between professional researchers and community leaders whose results were published in books and pamphlets and were widely disseminated.

The health metaphor then developed from a quest for treatment to one of prevention. Today, this metaphor is still going strong among many think tanks in developing countries that, greatly supported by a largely technocratic international development community, deploy the evidence-based mantra to defend their credibility and legitimacy in the policy space.

The Ford Foundation continues to play a critical role in promoting these ideas, but has been joined by new ones such as the Bill and Melinda Gates Foundation, which champions the power of science in the pursuit of social, economic, and political reform. Their emphasis on health and quantitative research methods is a testament to the permanence of this idea.

Efficiency and value for money

In the early 1900s, however, a new metaphor, which came back about a decade ago, took hold of the imagination of think tank pioneers and funders. The notion of efficiency was drawn from the world of physics and inspired the backers of the Twentieth Century Fund, the National Bureau of Economic Research, and the predecessors of the Brookings Institution. This idea assumes that just as machines could be made to work more efficiently by reducing wasteful loss of energy, so could public and social institutions. In the corporate world, efficiency had already been applied to labour (introducing time-keeping and marking the birth of professional management), production (in factories and offices), and finance (to reduce risk and maximise returns). It was just a matter of time until the idea was applied to policymaking.

Proponents of efficiency, according to Smith (1991), ushered in the introduction of a new relationship with both governments and the public: Gone were the associations of concerned and interested citizens, and in came the professional centres of experts and consultants. In this new model, the government, and by default the public, deferred to experts when it came to making informed decisions about matters of public interest.

In practice, the search for efficiency translated into an emphasis on quantitative research methods and a focus, above all, on the economics and finances of policymaking. We see the same happening today. Some think tanks in developing countries spend quite a lot of time not on the quest for new policy ideas but on the monitoring and tracking of aid flows and public funds. A Malawian economist working for a project managed by ODI once explained to me why few of his peers were able to work as economists: "We are only useful for tracking donor funds," he lamented. "Nobody is interested in funding new research on macroeconomics or economic theory more generally. At least not by us. It appears as if it has all been said." This reflects a view expressed by other think tanks and researchers who feel their work is limited to providing inputs to Northern researchers' own projects.

Combined with the health metaphor, the idea of efficiency has become a powerful driving force behind the recent embrace of the evidence-informed policy mantra and the results discourse that are all pervasive in the international development industry. As a consequence, some think tanks in developing countries have become (and in some cases have been set up as) not-for-profit consultancies, fixing and calibrating the system rather than thinking about alternatives. For example, according to Srivastava (2012), in Sri Lanka, "The decade of 1990s particularly saw entry of consultancy firms, advocacy groups, and associations in a big way that attracted foreign funding. A good example is the Centre for Poverty Analysis (founded in 2001), registered as a non-profit company to carry out an independent analysis of the causes, characteristics and impacts of poverty. It undertakes assignments for clients which typically include donors and international NGOs, and sometimes local NGOs and government." GRADE in Peru and Grupo FARO in Ecuador have very different structures but share the same contract think tank business model, and both have adopted governance arrangements to respond to research and project management contracting opportunities. Even ZIPAR, which benefited from a significant contribution from the Government of Zambia and the ACBF to conduct research at its inception, chose to set up a consulting team within the organisation.

Smith (1991) describes how the initial focus of these new centres had a local and urban focus: In New York, Henry Bruere and William H. Allen helped to set up the New York Bureau of Municipal Research (1907), which, although Tammany Hall[3] politicians dubbed the Bureau of Municipal Besmirch (poking fun at the authority that technocrats award themselves in a manner not dissimilar to the way that modern think tanks are made fun of), was frequently called upon by the heads of several city departments for advice.

The approach was scaled up in 1910 when President Taft set up a commission on economy and efficiency and named Frederick Cleveland of the New York Bureau of Municipal Research as its chair. Unfortunately for the commission, but fortunately for the future development of think tanks, it put forward proposals that political leaders found politically impossible to implement, and thus, Cleveland retreated to New York to develop a new strategy. In 1916, he set up the Institute for Government Research (which would later be expanded and renamed the Brookings Institution in 1927) and sought to influence policy from the "outside."

These new think tanks were different from other types of enterprises at the time, very much like many today are different from other civil society organisations. Unlike the municipal bureaus or the civil associations that, for the most part, depended on local businesses and the local

---

[3]A political organisation that acted as the Democratic Political machine in New York City.

community for funding and legitimacy, think tanks like the Institute for Government Research flourished thanks to the support of professional philanthropic foundations such as the Rockefeller Foundation. This was an entirely different arrangement, and the relationship between think tanks, the state, and society could be said to have changed forever.

To begin with, it freed the centres from the need for coordination between concerned but not necessarily professional policy entrepreneurs, experts, and funders. The approach also unleashed the capacity of the think tanks to deploy more scientifically robust quantitative methods that may not have been possible had they had to address non-technical public concerns.

In developing countries, most of the think tanks I have had the chance to work with draw their funding and support from the international development community, and so their relation to their fellow citizens is, at the very least, tenuous. Their only claim to legitimacy can be their intellectual credibility, which is taken by them to stem from their command of research-based evidence. To maintain this, they seek to influence the discourses and narratives within which policies are made and implemented. The promotion of evidence-based policy, in other words, has become an objective in itself.

By the way, the new strategy devised by Clevland worked: In 1921, President Harding signed the Budget and Accounting bill proposed by the Institute for Government Research. This marked an important new phase in the development of think tanks and the emergence of a new model of organisation.

Saved by the crisis and a new focus on planning for the future

The First World War brought many changes to the new modern think tanks that had been set up in the United States. The war, but particularly the experience of the war, had far-reaching effects on society, and think tanks shifted their attention towards them: The way in which the role of women in the workforce and politics transformed domestic life required a new system at home, and the collapse of peace on such a grand scale demanded a new international system to prevent all future conflict. This is what think tanks in Chile did in the 1970s and 1980s, according to Jeffrey Puryear (1994): focus on understanding the changes that traumatic events have in society. They too found and established their place in

Chilean politics by reflecting on the economic, social, and political changes that the country had undergone before and after the breakdown of democracy and the first decade of the military regime.

New regimes bring new challenges as well as opportunities. During the First World War itself, new government agencies were set up to deal with these challenges, and as the state became more complex so did the policy community that grew around it. The new agencies and committees demanded experts and, in turn, these experts went on to (often working in think tanks) propose the formation of new agencies and committees. Soon, the policy community became busy reflecting on the changes that had taken place and imagining new ones.

This focus on the past happened at a time when confidence in science was faltering: How could experts have not seen the Great War coming? However, a second crisis, the crash of 1929, boosted demand for their services.

Again, as in all crises, the Great Depression saw the rise of new interest groups converging onto Washington, D.C. in pursuit of political and economic influence. The same policy entrepreneurship that had kickstarted the process at the local level was now being unleashed at the federal level. This is not unique to the United States, but, as James McGann has argued, federalism certainly goes a long way towards explaining the abundance of think tanks there.

Instead of the scholarly investments that the NBER and the Brookings Institution had preferred, the Twentieth Century Fund (1922) saw a need for an entirely new approach. It invoked the old model of medical research of diagnosing and treating social ills, but this time advocated for turning ideas into big, actionable plans.

The 1930s hence ushered in the rise of the planning think tank. The Second World War and the increased demand for military planning capacity further strengthened the appeal of this idea. When the war was over, the social scientists, military researchers, and planners who had staffed the many agencies and bodies created to deal with the conflict remained in Washington and took on jobs in new forward-looking centres.

While the old research institutes focused their attention on how to avoid slipping back into an economic depression, new organisations began to plan for the future. The Committee for Economic Development (1942) took on a more proactive approach, for instance, focusing on finding what would be the proper balance between public and private involvement as the economy transitioned from war to peace. Its business model was an innovation — and simultaneously a return to the past.

The government also became more actively involved. Drawing from its wartime experience, a new way of accessing advice was devised: contracting. As a consequence, in 1948, the project "RAND on War Research" by the Douglas Aircraft Corporation became, with funds from the Ford Foundation, the RAND Corporation. RAND not only became a model for future generations of think tanks but it also played an important role in developing and popularising new analytical and planning methods such as systems thinking and the logical framework.

As a consequence, the think tank community grew to include organisations that served different masters: the public (whether this is directly or indirectly through the agency of charitable foundations and endowments), the corporate sector, and the government.

Salomon's House and the revolving door

In the years following the Second World War, yet another new metaphor emerged to explain the changing nature of the relationship between think tanks and policymakers: Salomon's house of expert advisors (Smith, 1991). This introduced a new model for think tanks to seek influence: the revolving door between them and the government. The economic crisis of the early 1930s presented an unparalleled opportunity for experts to influence future policy as political candidates saw the chance to break with the past. When President Roosevelt was elected, he drew from his campaign experts to staff his departments and the White House. Think tanks responded to the opportunity. The Brookings Institution offered free advice during the transition period, and experts from the NBER and the Russell Sage Foundation moved to Washington, D.C. to serve in emergency agencies set up in a rush to implement the complex set of new policies proposed by the New Deal.

In the United States, as in other countries today, expert advice was institutionalised. After 1946, new policies demanded that the president establish a Council of Economic Advisers. With more experts in positions of advice, the president needed more staff to manage them and provide a link between them and him. Slowly, the informal advisory roles played by, and the mechanisms available to, experts in think tanks were formalised in these new institutions.

Formal economic and social advisory bodies are now common in many developing countries — and could be considered internal government think tanks. For example, India has an Economic Advisory Council, Thailand a National Economic and Social Advisory Council, South Africa a National Economic and Development Labour advisory Council, Ghana an Economic Advisory Council, Nigeria a Presidential Advisory Council, and Namibia a Presidential Advisory Council, all of which report directly to key ministers, the cabinet, and the prime minister or the president. In China and Vietnam, the Chinese Academy of Social Sciences and the Vietnam Academy of Social Sciences play that very same role for their respective Communist Parties. They generate as well as draw ideas and advice from other think tanks and experts, both locally and internationally.

In the early 1960s, John F. Kennedy's administration brought into the government action intellectuals, *ivory tower activists*, or what James Smith labelled the Ministry of Talent (1991). Smith also reports that *The Economist* described Brookings' researchers as Kennedy's "experts on tap." This was more than a mere engagement with political issues. After years of close collaboration with politicians, experts had effectively moved into politics.

This same process oversaw the migration of researchers and experts from Chilean think tanks into the newly elected government of Patricio Aylwin in 1990. Throughout the 1980s, these think tanks had developed plans for the new democratic government, while also working closely with politicians of the opposition. Jeffrey Puryear (1994) provides a clear account of the extent of this migration of experts into politics: The members of the new cabinet included Ricardo Lagos (Ministry of Economy), Alejandro Foxley (Ministry of Finance), Edgardo Boeringer (Minister of the General Presidential Staff and Chief Political Strategist), René Cortazar (Ministry of Labour), Carlos Ominami (Minister of Economic Affairs), Germán Correa (Ministry of Transport), Francisco Cumplido (Ministry of Justice), and José Antonio Viera-Gallo (President of the Chamber of Deputies). All had been active public intellectuals and researchers in those think tanks that were set up in the two decades prior to the formation of Aylwin's government, who then found themselves as chief providers of ideas and architects of policy.

Inevitably, experts inside the government looked for experts outside of it. In Kennedy's United States, researchers from the Brookings Institution and RAND became very close to the government and its policies. While Brookings supplied the White House with experts, RAND was the main recruiting ground for Robert McNamara's Department of Defense.

Just as in the United States, the presence of experts in the inner circles of politics and high-level policy spaces changed the political debate in Chile. Disagreement stemmed from technical rather than political rifts, and progressively the power of technocrats over ideologues prevailed (Puryear, 1994). This became a trend that would continue for some time in the history of the modern think tanks and that, as we will see in future sections, marks a critical moment in the emergence of think tanks in developing countries.

The presence of experts as part of the policy community further changed political discourse. It strengthened the legitimacy of technocrats in policymaking and reinforced the relations between governments and experts in think tanks. The links between what happened in the United States and Chile are not coincidental. As I will describe later on, dynamics that played such a critical role in the early stages of the development of think tanks in the United States were also responsible for developing Chilean research capacity in the 1950s and 1960s and their think tanks in the 1970s and 1980s. Furthermore, the geopolitical and economic ideas that were being developed in the United States at this time were soon to impact Chilean politics and policymaking by means of the links being forged between intellectuals and experts in both countries.

The ideological marketplace

The 1960s introduced another twist in the history of modern think tanks (Rich, 2006). The Hudson Institute (1961), as well as focusing on planning for the future, was explicitly ideological (a commitment to free markets and individual responsibility), and its research reflected pre-formed values. This constituted a break with the conventions of neutrality and academic objectivity. Concerns about this new approach led, in the late 1960s and early 1970s, to a backlash against the influence of think tanks in politics, which, in a context of significant reductions in federal funding, particularly that available from the Ministry of Defense, had important effects on the think tank community. Furthermore, new tax laws limited the support given to think tanks and their activities in policy advocacy or influence, and as a consequence, organisations like the Ford Foundation (which had hitherto been the largest funder of think tanks) began to reduce their funding towards them.

Think tanks had to seek new sources of funding. The prospect of attracting wealthy patrons encouraged the development of new fundraising skills and more overtly ideological policy messages than in the past. This was met by a positive response from those concerned with the perceived permissive moral liberalism in American politics and society of the 1960s: conservative businesses and corporations, neo-conservative groups, Christian groups, and neo-classical economists. A new model of think tank developed, this time founded not by philanthropists or scholars looking for new homes for their trade, but by political insiders: The Heritage Foundation (1973) and the Cato Institute (1977), two of today's most influential conservative think tanks in the United States, were set up by people already working in politics and who understood think tanks more as part of the political arsenal of conservatism than simple suppliers of ideas. This explicit alignment led to a sort of an intellectual arms race with their liberal counterparts (Bai, 2008).

These think tanks offered above all access to political power and joined an existing community of academic and contract organisations more accustomed to private scholarly advice than public engagement and debate. To compete, they too had to adopt new practices and skills, and a new metaphor, arguably the most powerful in the international development discourse on think tanks, was born: the marketplace.

It is difficult to think of a study or event involving think tanks in which the term, "marketplace of ideas" is not used. Even in China, at the First International Symposium on Think Tanks organised by a group of think tank scholars in June 2013, the marketplace was a feature and the focus of much of the discussion. But where does it come from?

The marketplace metaphor was developed by the incorporation of marketing practices and expertise into the world of think tanks in the 1980s, according to both Rich (2006) and Smith (1991). Encouraged by their new sources of patronage, think tanks changed their structures and activities. For instance, the American Economic Institute launched the American Enterprise Magazine, while the Brookings Institution launched a new series of policy briefs in 1996 and followed this by appointing a vice president for communications.

These changes were resisted, particularly by the staff, as they inevitably implied changes to the way they were hired and managed within the centres. Brookings, for example, changed the practice of automatically renewable contracts: Richard Haass (Vice President and Director of Foreign Policy Studies at Brookings) commented that he wanted to shed the notion that the think tank was All Soul's College, a graduate college at Oxford University often associated with the idea of the ivory tower (Rich, 2006). According to Rich, Haass suggested that scholars at Brookings should serve for shorter periods, sufficient to reflect and write, and then return to public service. The same resistance is seen today across think tanks that begin to adopt new approaches to engaging and communicating with policymakers.

In the late 1990s and the first decade of the 21st Century, think tanks' capacity to market their ideas extended beyond their traditional boundaries. The access to new and cheaper technologies, and the weakened state of traditional (print) media and political parties, has allowed (or arguably forced) think tanks to develop communication outputs and channels that have significantly increased their influence in the public space and even appropriated functions of other policy players (Tanner, 2012; Scott, 2011). For instance, not only have they become the source of regular commentary in traditional media outlets but they have also taken on the role of broadcasters themselves — producing and disseminating their own content directly. In international development and in developing countries, think tanks have taken to the implementation of projects, thus operating as consulting firms, producing content for global and local media, and directly drafting and implementing policy, either commissioned by donors or governments.

The marketplace metaphor is one of great importance for think tanks across the developing world. Among them and in their interactions with their funders and the organisations contracted to support them, lies a potential misunderstanding. There is a difference between appealing to the metaphor of a marketplace of ideas and promoting a marketplace for think tanks.

The former seeks to describe the manner in which think tanks compete with each other with alternative ideas (for example, explanations of problems and solutions). The latter encourages competition for funding and staffing that, while positive in some respects, can also lead to the commodification of knowledge and the privatisation of research and advice. And this, as the pluralist and elite perspectives make clear, is detrimental to intellectual plurality and, ultimately, democratic values.

The Great Conversation: More of the same or an original idea?

A fifth metaphor, described normatively by David Ricci as a kind of utopic style of politics — in which think tanks would play a key role — is that of a public conversation (Ricci, 1993). Ricci labelled it the Great Conversation, invoking an image of policymaking as "open," with think tanks playing an important role in mediating and improving the breadth and quality of the debate.

In a context in which public policy is increasingly carried out in a transparent and accountable manner — often by adopting new technologies — and thus disrupting more traditional, and exclusive, communication channels between think tanks and governments, the idea of an inclusive (and great) conversation is appealing. Think tank models that draw their strength from their networks or associates have emerged.

Stephen Yeo, CEO of the Centre for Economic Policy Research, has pioneered the idea of a think-net (Yeo and Portes, 2001). Think-nets draw their experts from other organisations (for example, economists based at European universities in the case of the Centre for Economic Policy Research) and are therefore able to take advantage of the investments made by these in their staff. Equally, they are capable of reaching, through their associated experts, policy spaces they would not have been able to if they had acted on their own. By relying on associates, CEPR, which is based in London, is able to access expertise from and reach policy spaces across Europe. A similar model has been effectively adopted by the Centre for Global Development through its visiting and non-residential fellows. Other think tanks in developing countries are following a similar strategy: Grupo FARO in Ecuador launched a new associate researchers pro-gramme in early 2012 to make more use of its research networks; ZIPAR in Zambia has associates at the core of its business model; Gateway House in India relies on associates to undertake important new studies; and the Instituto de Estudios Peruanos's entire senior research body is also associ-ated with national and international universities.

Researchers and think tanks are also coming together through loosely defined associations and networks. VoxEU (an online initiative from the Centre for Economic Policy Research) is a perfect example of what Nick Scott calls the digital disruption of think tanks' traditional spaces (Scott, 2011). By convening leading economic experts, VoxEU has become a source of focus for both producers and users of economic policy research and analysis. Since it is far more popular than any other online space, think tanks interested in engaging in economic policy debates would do better to focus their efforts on communicating through VoxEU instead of attracting audiences to their own websites.

At the other end of the policy community, Guerrilla Policy was a short-lived British think tank set up to attempt to work on policy solutions from the bottom-up. It sought to encourage policy practitioners (nurses, teachers, civil servants, etc.) to work on policy solutions for problems and challenges that affected them. Digital tools and social networks have made their involvement possible.

Grassroots think tanks like foraus and student-led think tanks have played a central role in this phase of the history of think tanks.

It is still early to make sweeping statements about the way in which this digital disruption has affected think tanks and whether there is a distinct new model out there. Claims that the old model of think tanks has passed its sell-by date have been made since 2011 by many political commentators in the British press.[4] It does appear, however, that there is a sense that a new way of thinking of think tanks is necessary.

The introduction of AI to the think tank space has certainly upended many plans and expectations.

It is also clear that over the years think tanks have become more and more a permanent feature of the policy community rather than external and independent catalysts of change. They are insiders more than outsiders. Heritage Foundation's Project 2025 played a central role in the 2024 U.S. Presidential elections.

As funding and resources have become more dependent on fewer sources, and as these and their main promoters have also become more closely linked to the political space, they have turned into more opaque organisations and less representative of wider professional, economic, or social communities. The idea that evidence should play a crucial role in policymaking may not have changed much, instead simply metamorphosing over the years to accommodate changes in the political context and the advent of new technologies and relations. However, the role that citizens have played in pursuing this idea has changed: from participants to observers.

---

[4] See, for example, George Monbiot's demand for transparency or Andy Williamson questioning of think tank's quality, both for *The Guardian* newspaper.

The Great Conversation metaphor is hence less a description of the current model but more a reaction to what think tanks have become. Taking the evidence-based policy mantra to an extreme has, in some cases, dehumanised and depoliticised the development process. In my view, think tanks, their funders, and the public have become more concerned with objectives, methods, and evidence than with their environment and the societies they serve.

New technologies, particularly those that allow the public to participate more actively in the development of public policy debates, are changing the roles that think tanks must play in their societies. Increasingly, they are finding that ideas and evidence are no longer their sole domain and that decision-makers seek them out for guidance and to make sense of the conversation.

The Great Fight

Unfortunately, the Great Conversation narrative has been replaced by something more aggressive and divisive. Think tanks across the world are reporting a growing political polarisation and division (On Think Tanks, 2024). This is promoting responses from funders and think tanks alike that pit the community against each other.

Throughout this book, I hope to show that think tanks are easily used to promote and pursue others' interests. Recognising themselves as part of a unique community — even considering all the organisational and contextual differences that set them apart — offers the best chance to fight back against this weaponisation of think tanks.

Changing Relationships

This history of think tanks highlights a number of interesting characteristics present in many centres today. An emphasis on academic research, delivering projects and programmes on behalf of public and private bodies, the use of marketing tools to communicate their findings and brands, and developing informal political relations are common features among think tanks, and these owe their current state to a long and complex history of interrelated and competing ideas.

Particularly interesting, however, is the emergence of a history of changing relationships between funders, citizens, researchers or experts, and policymakers. Originally, citizens, funders, experts, and even policy-makers appeared to come together around think tanks to address problems in their communities. The appearance of large funds and the professionalisation of the social sciences "closed" the door to the average citizen and limited his or her role to that of philanthropists or the public.

The similar professionalisation of the government had an effect in the same direction. And the incorporation of marketing ideas further transformed the public and policymakers into simple audiences or clients. In essence, advice over issues of public interest has become increasingly private, and its provision is subject to the rules of the market.

Something akin to this can be found in developing countries where funding for economic and social research is still the prerogative of foreign funders. Policymakers are described as target audiences or the demand side, think tanks as suppliers of advice, and citizens are excluded from participating in evidence-informed policymaking. That remains, it appears, the constituency of NGOs and other civil society organisations.

It must be noted, however, that the metaphors have, in many cases, simply piled on top of each other. It is not uncommon to hear think tank directors use several of them in any presentation of their organisations. Alison Evans's description of ODI as "a think tank with a twist" was a reference to the organisation's combination of business models but also, although not explicitly, acknowledging the various ideas that drive it. Labels like "think and do tank" are equally unhelpful.

Finally, this history is as much about the changes in the way in which think tanks promoted their ideas as it is about the promotion of the idea of evidence-based policy. This has served both to encourage more informed policymaking as well as to guarantee (or argue for) think tanks' legitimate role in politics.

Chapter 5

A More Complex Account: Case Studies

While the history of think tanks in the United States can shed some light on the drivers of modern ideas of think tanks in the developing world, it cannot be expected to comprehensively explain the formation and development of think tanks everywhere. The cases presented in this chapter are also not necessarily comparable and no single framework is being put forward at this time. Instead, they provide an opportunity to look into and reflect on the complex lives of think tanks in developing countries in a manner that takes us beyond the organisations themselves. The case of Chile's think tanks describes how they were set up outside the State as a vehicle for intellectual and political resistance to the technocratic and authoritarian military government of Augusto Pinochet. In Indonesia, the history of the SMERU Research Institute provides an illustration of a common origin for many think tanks in developing countries: as a spin-off, in this case, of a World Bank project. The same diversity in origin can be found among think tanks in Peru, whose affiliations are still closely linked to academia, the private sector, and civil society organisations or NGOs. In Zambia, the State, donors, and NGOs have set up think tanks, and now, political parties as well. Peru and Zambia offer examples of the diversity of think tanks that can be found even within the same country.

This section helps to draw out some initial emerging issues that resonate with the broader global literature on think tanks.

I have left the case of Chinese think tanks to Chapter 7 because it helps to make an important case about the role that even single-party governments can play in the development of a strong think tank community. It is worth keeping it in mind.

Specific Circumstances in Think Tanks' Origins and Their Development

The origin of think tanks can be rather circumstantial and opportunistic. We often assume that they are organisations that demand a great deal of planning and up-front costs before they can be set up and start functioning. This is something I have heard from many researchers and donors when confronted with the question of why they do not set up and fund new think tanks in countries where there are only a handful of them, and yet there is demand and funds for more. This is not to say that planning (this is in fact a feature in the Belgrade Centre for Security Policy's story) does not play a part in their formation, but rather that the real forces behind their origins have often little to do with grand purposeful strategies of policy change or institutional development.

Neither do there seem to be any clear rules — or where there are broad rules, there are also exceptions. Their emergence appears to be more a matter of rebellion, of researchers, policy entrepreneurs, philanthropists, and international funders bucking the trend, breaking the rules and taking advantage of opportunities.

Throughout Latin America, for example, many think tanks were set up simply as a means for academics to complement their income with research consultancies and compensate for the meagre salaries they received from public universities. The same is true in Asia and Africa, where many of today's most successful think tanks are staffed by academics who have stopped teaching altogether or retain limited pedagogical responsibilities. In Africa and parts of Asia, as we shall see in the case of SMERU in Indonesia, and CTPD and ZIPAR in Zambia, I argue that donors and non-governmental organisations, respectively, originally set them up as implementing vehicles for their own policies — although, to varying degrees, the organisations have been able to steer away from this origin. But even in there, one can find think tanks driven by individuals, parties, and civil society organisations in search of alternative means of influencing or monitoring policy and policymakers. In Chile, local versions of the same metaphors have driven the rise of modern think tanks.

The stories of Chilean think tanks and SMERU in Indonesia, presented here in greater detail, are particularly relevant to this point. While they are very different cases, they offer an excellent account of two interesting and insightful origins of two particular types of think tanks: ideological and technocratic. The origin of think tanks in Peru described in this chapter, supports the idea that think tank formation can be and often is rather opportunistic. It further contributes by showing the roles that academic, corporate, and civil society forces play in their origins. This is echoed by BCSP's story.

Chile: Spaces for survival

The history of Chile's modern think tanks, carefully described by Jeffrey Puryear (1994), emphasises the importance that certain political events can have on the policy landscape and the formation of think tanks, as well as an indication of how political openness may not always be, as is sometimes reported (Bellettini and Carrión, 2009), responsible for their emergence. It presents, too, an interesting parallel to the story of think tanks in the United States described in the previous chapter.

In Latin America, the concept of think tanks, and certainly the label, has only recently gained notoriety but many organisations have been performing think tank functions as early as 1779 when the Sociedad Académica de Amantes del País met in Peru to exchange ideas about an independent nation and published a journal, *El Mercurio Peruano*, to promote its vision.[1] Colombian think tanks developed just over half a century later during their own struggle with the formation of the modern Colombian Republic and have evolved in close relation to its political parties and the media (Mendizabal and Sample, 2009).

Chile, however, is the country most often associated with think tanks in Latin America. Chilean think tanks are probably the closest one may get to think tanks that reflect the Anglo-American think tank ideal. Nonetheless, their story, or, more appropriately, their history, is nothing like that of the rise of think tanks in the United States, even though it is closely linked to it.

When Augusto Pinochet staged a coup d'état against the government of Salvador Allende on 11 September 1973, many academics found themselves on the front lines of State repression against intellectual and ideological dissidence. Following the coup, think tanks offered, quite literally, safe havens for researchers and their ideas. The Pinochet coup was,

---

[1] The Sociedad Académica de Amantes del País was inspired by the learned societies common in Europe at the time. The Royal Society of Arts (The RSA) in the United Kingdom remains the oldest, in existence, think tank of its kind.

however, only the trigger for the formation of think tanks: The founda-tions for this community had been laid down many years before.

Historically, Chile has had one of the best education systems in Latin America and is one of the most politically sophisticated and mature coun-tries in the region (Garcé, 2009; Cociña and Toro, 2009; Puryear, 1994). Not surprisingly, it was a favourite of the international developed com-munity: It presented democratic politics, a strong, effective state, was open to new ideas, and was an alternative to Cuba's socialist promises.

For decades before the coup, Chilean universities had benefited from generous funding from foundations in the United States and Europe, asso-ciations with prestigious universities, and the visits and collaboration of academics from various developed nations who helped to build one of the most developed academic communities in the developing world. In 1960, Chile's adult literacy rate reached 84% and Chilean educators were in regular contact with their European and American counterparts. The Chilean intellectual scene was also particularly interesting. Unlike other Latin American countries, both academics and politicians were interested and committed to an open exchange of ideas on issues of public interest. This relationship is clear in the decisions that helped to develop Chilean academic and research capacity.

According to Jeffrey Puryear's detailed account of the history of think tanks in Chile (1994), Juan Gómes Millas, the Rector of the Universidad de Chile, sparked a systematic effort to institutionalise scientific research in Chilean universities, which was met by public interest and funding. In 1954, the Chilean government channelled 0.5% of all non-municipal taxes, as well as all customs and export duties, into a University Construction and Research Fund and established the Science and Technology Committee to advise the government on national policy.

In 1958, Roger Vekemans Van Cauwelaert, a Belgian Jesuit, estab-lished a School of Sociology at the Universidad Católica, which later counted on the support of the Ford Foundation. Also in 1958, Gómes Millas set up the Institute of Sociological Research at the Universidad de Chile, and, at about the same time, the Facultad Latino Americana de Ciencias Sociales (FLACSO) and the Latin American School of Sociology (ELAS) were founded with two foreign directors from Spain and Switzerland, respectively (Puryear, 1994).

This rapid development in sociology was complemented by develop-ments in economics. Argentine economist Raul Prebisch inspired the formation of ECLA (now known as the United Nations' Economic

Commission for Latin America and the Caribbean, CEPAL, from its name in Spanish) in Chile. In 1957, Latin America's first graduate training programme in economics, ESCO-LATINA, was established, and soon a new generation of economists followed: The number of economists trained in Chile grew sixfold between 1960 and 1970 (from 121 to 727). Half a dozen new research and teaching institutes had been established in the two main universities (Universidad de Chile and Universidad Católica), and this was complemented by the presence in Chile of well-trained foreign social scientists working in Chilean universities and international bodies (Puryear, 1994).

Donors met this national commitment with great enthusiasm, made additional contributions, and encouraged further exchanges with American and European universities. By 1970, there were eight public and private universities highly subsidised by the government and enrolment in 1973 had doubled since 1969 to 147,000 (Puryear, 1994). Universities were thriving.

The effect this investment had was that by the 1970s, Chilean intellectuals constituted a strong and active community stretching across academia, civil society, the media, and politics. They had access to the country's elites through the institutions of the media, political parties, and professional schools. They enjoyed good links to their professional peers in Europe and the United States; some worked in international organisations, others in private research organisations, and many in the government.

But this community, as Puryear describes it, was dominated by a kind of intellectual who emphasised "society as a whole" rather than as a collection of separate parts, divided into specialised policy areas or sectors. In a way, this constituted a stance against experts or technocrats and the prevailing metaphors (all variations of the evidence-based policy mantra, as we saw in the previous chapter) that had driven the initial formation of think tanks and still does so in developing countries today.

Hence, the Chilean intellectuals were characterised by their strong orientation towards discursive-level ideological dialogue and an active distancing from instrumental or specialised research. In short, it wasn't demand-driven, as Srivastava laments that most research produced in South Asia has turned out to be (Srivastava, 2012).

The new regime brought with it a reaction to this political culture. Pinochet's government was anti-politics — it replaced politics with administration and politicians with technocrats. In this new context, two institutions rose to dominate Chile's public life: the Armed Forces and the

Catholic Church. The Armed Forces represented the technocratic pragmatism of the new regime that distanced itself from dialogue in search of practical and more immediate solutions to real-world problems. It reflected the metaphors of efficiency and planning that had influenced the development of think tanks in the United States. On the other hand, the Church, possibly influenced and motivated by its own discourse of values and society, which was not too dissimilar to that adopted by public intellectuals before the coup, became a convenor for the opposition. As the only institution that was able to hold public meetings without first seeking approval from the regime, it became the host of committees, centres, associations, and other gatherings that were to develop into Chile's modern think tanks community (Puryear, 1994).

While politicians were being persecuted, intellectuals were relatively free to work as long as they focused on more technical matters. Academic analysis was not against the law: politics was. As a consequence, the focus on society as a whole was replaced by more technocratic concerns. Additionally, with universities under attack, scholars realised that they needed a new home. At the time, there were a few international organisations (FLACSO and CEPAL, for example) and private research centres, but these were not enough for all those looking for new homes.

New types of organisations were needed: The Chilean Institute of Humanistic Studies (ICHEH), supported by the Konrad Adenauer Foundation, was set up in 1974. Although initially just a publishing house, ICHEH was established within the Church's legal framework and hosted by it. In turn, ICHEH established the Centre for Socio-Economic Research (CISEC), led by a Jesuit intellectual, Mario Zañatu, to produce reliable data and analysis on topics of national interest. CISEC published regular reports, officially intended for the Bishop's Conference but of clear general public interest as well.

ICHEH and CISEC were not initially meant to replace the older institutions. They were emergency measures taken by academics, the Church, and other intellectuals who, at the time, would have expected a return to "normality" in the medium term. This sense of emergency is reflected in the way the Church, right after founding ICHEH a year earlier, set up the Academy of Christian Humanism in 1975 as an institutional umbrella for academic groups cut adrift from traditional institutions. Funded by the Ford Foundation, the Academy grew into six research programmes employing over 200 researchers. It promoted study circles that stepped in for professional schools. In fact, much of the work of these new think tanks focused on the formation and facilitation of spaces for dialogue and reflection.

Eventually, and as it became clear that this new state of affairs was in fact transforming Chile, many autonomous centres emerged by the end of the decade: for instance, the Latin American Corporation for Economic Studies (CIEPLAN), the Centre for Social and Economic Studies (VECTOR), the Centre for Social and Education Studies (SUR), the Latin American Centre for Political Economy Studies (CLEPI), and the Development Studies Centre (CED).

An organisation given particular attention by Pruyear's account (1994) is the Group of 24 (or Group of Constitutional Studies), which was set up in 1978 by Christian Democratic politicians and lawyers to debate the regime's plans to establish a new constitution. The Group of 24 is important because the dialogue that took place within it led to the formation of the political Democratic Alliance in 1983, the National Accord for the Transition to Full Democracy in 1985, and the 13-Party Concertación that engineered the victory over Pinochet in 1988. The impact that think tanks had on Chilean politics and policies cannot be clearer — and this is, in my view, one of the best examples of the contribution that think tanks can make to a society found in the literature.

These new organisations had distinct and new characteristics that marked the appearance of modern think tanks in Chile, and possibly in Latin America.

First of all, they became more numerous and were more likely to be organised around specific policy issues than entire academic disciplines, somewhat reflecting the more technocratic approach to government that they faced. In this sense, the centres oriented their work towards real-world problems and developed closer ties with unions, labour leaders, and community groups. This active concern with the public interest is a unique characteristic of modern think tanks and what sets them apart from other organisations that focus on the research of economic, political, or social issues from a more disinterested or distanced position, such as academic centres or even research consultancies.

Ironically, researchers and intellectuals in the new think tanks found a new sense of freedom and were less constrained than before to collaborate with non-academic groups and particularly with politicians. Additionally, the think tanks they founded and helped to develop were relatively more politically cohesive than the broader academic departments had been in the past. In other words, researchers came together around particular ideas or policy positions. This helped, no doubt, to develop some of these think tanks' brands.

These organisations were also more aware and adept at communications. This led to the formation of a social communication network by means of events and meetings. These spaces offered an opportunity to develop and voice a running commentary to the regime's own policymaking (itself heavily influenced by research and the now famous Chicago Boys, whose ideas have had a long-lasting effect across Latin America and the developing world), as well as, quite literally, a safe space for policy debate.

Most importantly, the centres attracted significant support from foreign academics and academic institutions that offered fellowships and invitations for Chilean researchers to speak at their events. But as Puryear points out, the involvement of foreign donors, many of whom were protecting their long-term investments, meant that researchers had to focus much more on hard facts rather than ideological reflection, and were subjected to new ways of working, with shorter time frames and deadlines. This reinforced the new primacy of technical knowledge over values.

Finally, these think tanks adopted unique roles or functions. According to Cociña and Toro, they performed the following functions (2009: 101):

- Provided safe spaces for opposition intellectuals and academics.
- Sought to interpret and understand the changes imposed by the new regime, as well as the causes that led to the breakdown of democracy.
- Helped to develop the skills and technical knowledge of future policymakers.
- Generated spaces for engagement and dialogue between the opposition parties and even those closer to the government.
- Developed the ideas and plans for the future democratic government.

This was, of course, not the end of the story. As Matías Cociña and Sergio Toro (2009) have shown, after their consolidation during the 1970s and 1980s, Chilean think tanks followed a path of political Darwinism — borrowing Michael Coppedge's term for the evolution of political parties in Latin America. The 1990s, with the new coalition government in place, saw the rise of think tanks associated with the Pinochet regime — eager to protect some of the economic reforms that had been adopted in Chile. The coalition itself weakened the very think tanks that had supported its rise to power by drawing senior researchers and leaders from them to staff the government. Political consensus further reduced opportunities for intellectual renovation, and many think tanks came close to disappearing. As a consequence, new forms of organisations developed: across party networks and associations, for instance (Cociña and Toro, 2009).

The recent history of Chilean think tanks raises several issues that merit our attention and that will help explain other cases. Firstly, there were already independent research centres in Chile, albeit in a reduced form. Chile's intellectuals were also particularly well connected to an international network of peers working in universities and think tanks in the United States and Europe. Therefore, the modern think tank model was, at the very least, familiar to researchers, policy entrepreneurs, and funders in Chile.

Chilean research and policy capacities, characterised by the quality of its academic and political institutions, were high. The collapse of the democratic order, while devastating to Chile's political system, did not entirely dismantle the policymaking apparatus and certainly did not reject the role that knowledge could play in politics. This offers hope to other parts of the world where democratic institutions are under threat today and suggests a role for think tanks.

Expertise was still in high demand during the Pinochet regime, but a new model for using knowledge had emerged: one that very much copied what was already popular in the United States at the time, where think tanks were acting as advisors and consultants to policymakers. This demand explains the existence of think tanks in non-democratic regimes like China or Saudi Arabia.

Nonetheless, the coup d'état did come as a shock and it had important effects on the work that researchers took on. Chile had a long democratic tradition and the new order took many by surprise. This meant that even before attempting to engage with the new regime, Chilean intellectuals were faced with more fundamental questions regarding the causes of the democratic breakdown and the manner in which Chilean politics, economics, and society were changing. The work of the think tanks during their first years focused on these, very much so as think tanks had done in the United States after the Great War, the Great Depression, and the Second World War, before turning their attention to the future.

The support of the international community was also crucial, but its success is certainly down to its funds landing on fertile ground. Particularly in the 1980s, when many of the new think tanks had consolidated and turned their agendas to more immediate policy concerns and the long-term objective of reinstating democratic order, European and American foundations provided important resources for the production of research and the organisation of events. International patronage strengthened their legitimacy before a regime that sought to present itself as progressive and open to the world.

In a way, then, these new think tanks were not established with specific policy objectives or research agendas in mind; they emerged quite simply to host and protect the intellectual human capital that had developed in Chile over more than two decades of continuous domestic and international support. The modern think tank model offered researchers a space to regroup and make sense of the situation Chile was living through and learn how to engage in the new world that was being built. This meant relearning their trade and refocusing their attention on more tangible real-world problems and, dare I say, embracing the technocratic model put forward by the regime.

It is important to stress this point because it may help explain the success of many think tanks over time and the relatively large number of them that were set up at the time. The Chilean case presents an alternative (or at least complementary) to the narrow policy-influencing mission. Think tanks were conceived (at least in the beginning) as a home for intellectual activity: as a space for continuity rather than as agents of change.

Also important is that it would be impossible to talk of a single Chilean tradition of think tanks. The model that worked in Chile changed dramatically from the pre-Pinochet days to the present day. At different stages, and presented by several hurdles, think tank promoters had to adapt the model to new conditions. Ironically, their success was rewarded by an unintentional raid on their staff as the new democratic government sought to fill their ministerial positions and thus threatened to bring many centres' existence to an end. It is clear, then, that think tanks cannot expect to fulfil the same functions or pursue the same objectives indefinitely. As a consequence of Chilean think tanks' constant response to their changing environment, several models remain in existence today. This fluidity in definition, functions, and structure is a characteristic of a healthy think tank community.

Chile's think tank history is also a reminder that such long and rich accounts exist beyond the United States. It is also driven by strong ideas of the role that research and researchers play in a society. Many of these ideas are shared across contexts — even if they are adapted to each reality.

Their study should be mandatory for any attempt to work with or support think tanks in developing countries, particularly if such effort originates from abroad and has the potential to affect a country's political culture and values.

Fluidity: The Belgrade Centre for Security Policy (BCSP)

The BCSP offers an interesting opportunity to revisit the issue of the labelling of think tanks, discussed in Chapter 2, and provides further illustration of fluidity in think tanks' development. A key element of the definition adopted was the self-identification of the organisation as such. Medvetz's view that the act of identification is a political one presents a very powerful argument for awarding think tanks the right to do so (Medvetz, 2008).

I visited BCSP in April 2012 to conduct a review for one of its funders, the Think Tank Fund of the Open Society Foundations. Among those whom I interviewed in relation to BCSP were several representatives of other civil society organisations, some of which were also in the process of adopting the think tank label. And it was in reflecting on this that this account came about. BCSP's use of the label was rather new. It started out as an activist NGO called the Centre for Civic–Military Relations in the late 1990s during the days of the former Yugoslavia and the rise of the opposition to Miroslav Milosevic. Like many of the civil society organisations at the time, it was, in essence, an organisation motivated by the belief in democratic values. Its founders, however, had, and still have, close ties to academia, and so the centre evolved from an NGO to a research institute (it founded the Belgrade School of Security Studies as a project of the centre), and finally into a think tank.

The decision to become a think tank was not taken lightly. After the fall of Milosevic and the rise of a new political class that had emerged from the ranks of the civil society organisations that opposed his regime, the need for activism in the security sector changed. The centre's initial efforts had focused on opening up a space for public debate on security policy and the role of civilians in the military. When the political space opened up, BCSP's leaders recognised that changes in its strategy and focus were necessary and hence turned towards developing a new identity as a research institute and the formation of new cadres of civilian experts.

Becoming an academic institute would not only confer on it the credibility that is awarded to academics in Serbian policy communities but also access to public funds earmarked for scientific research. The choice, even for an organisation comfortably funded by foreign sources at the time, was made, conscious of the role it sought to play in its own political space and of a sound reading of current and future challenges.

The decision, later on, to become a think tank was equally contextual. The funds available from the state proved not to be sufficient for the effort that becoming a research institute would demand. At the same time, the field of security studies in Serbia matured and several faculties began to offer undergraduate and postgraduate degrees. In fact, many of BCSP's former and current researchers work in academic departments and focus their efforts on the production of academic literature on the sector.

Furthermore, its initial investments to become a research institute had prepared it for the next phase. In 1996, the centre had set up the Belgrade School of Security Studies, in essence an in-house graduate school for a group of young civilian researchers. This group later formed the core of the think tank's research staff.

The think tank label is, of course, an external one, but the choice was local and heavily influenced by the centre's new management and its links to initiatives such as the Think Tank Fund. The change in labelling at BCSP was accompanied by significant changes in its functions and structure: The founding leaders of the centre retired to give way to a new generation of civilian experts. They established an entire programme to build the centre's own research capacity. The new leaders rebranded the centre and changed its name to reflect its broader objectives, and, more recently, BCSP employed new senior management and communication staff to ensure that the organisation can live up to the expectations of the think tank label.

Becoming a think tank has also meant a shift in the balance between research, advocacy, and training, which remain BCSP's main approaches to influence. Research became more instrumental and concerned with real and current policy challenges; it is no longer necessary to introduce a new idea or convince policymakers or the public of the existence of a problem. Their advocacy work has evolved from activism to more structured research communication channels and tactics that embrace constructive engagement. Finally, its training and education efforts are increasingly geared towards the development of new competencies and skills among future and current policymakers and practitioners in collaboration with established academic institutions.

The other organisations that I encountered in Belgrade, like many in other developing countries, appeared to have been driven more by a reading of their funding environment rather than of their broader internal and external contexts. Therefore, the relabelling exercise that some are undertaking is inspired, instead, by the promise of new institutional funding, much like the type that they received in their early years, or new contracts available from an international community increasingly interested in think tanks.

This is by no means a commentary on these organisations' competencies. In fact, it was my impression that Serbian politics still needed centres like the Belgrade Centre for Human Rights, the European Movement in Serbia, or the Centre for Euro-Atlantic Studies. While they are not think tanks in the stricter sense that BCSP is, they play equally important roles — roles that, when carried alongside BCSP's, can lead to greater impact.

Why were these centres, including BCSP, becoming think tanks? Like most countries on the list to become members of the European Union (or, as in the case of Serbia, on the list to be considered to be on the list), Serbia was facing important changes to the way in which the international community funded civil society. Several bilateral funders had left or were in the process of withdrawing, targeting their funds to poorer countries or issues that are more closely associated with the plights of the poor.

Furthermore, new funds such as those of the European Union to its neighbours were increasingly channelled through official channels. These funders, as in other parts of the developing or emerging world, were by and large no longer interested in building new institutions (it is their view that this has been done already) and were instead focused on the delivery of timely advice to policymakers.

Faced with an increased competition for aid funds, networks, social movements, interest groups, activist NGOs, and consultancies, many of which are described by other policy players in the government and the media as "one-man/woman shows," began begun to label themselves as think tanks as a way of differentiating themselves from the rest. At first sight, this relabelling makes sense. It could be driven by the same reading of the context that led BCSP to change its identity in the mid-2000s. But this is not necessarily the case.

It could be, however, that in searching for new foreign funds, Serbian civil society could label itself out of existence.

Since my visit, the context in Serbia and the Western Balkans has rapidly deteriorated. Democratic institutions have been weakened across the region. Think tank leaders have had to make difficult strategic choices to survive.

In a series of interviews with former and current think tank leaders from the region, we documented these choices and the evolution of the sector. In Serbia, in particular, the space for think tanks to operate has closed. The context has forced BCSP to revisit old functions of watchdogs and advocate for fundamental rights at the expense of the policy research capacities it had developed.

The early years of BCSP's history were strongly influenced by the role played by the towering figure of Professor Miroslav Hadzic, one of the centre's founders. Even in 2012, several years after retiring from BCSP's direction (although he did so to become an active member of the board), he was closely associated with the centre. But unlike other founder-directors, Professor Hadzic took a step back and handed over the reins to a new generation of policy entrepreneurs. His role, as well as that of other founding members, changed to focus on supporting and mentoring the organisation and its researchers. The new leadership, particularly, BCSP's former director Sonja Stojanovic, was therefore able to adopt new ideas related to the future path that the organisation could take.

Arguably, this ingrained flexibility and adaptability into BCSP's DNA.

BCSP's story illustrates the changing nature of think tanks themselves. Not only do we find different kinds of think tanks in a country but we can also see that the same organisations can go through changes as well.

The SMERU Research Institute (SMERU) in Indonesia

SMERU's former director, Sudarno Sumarto, outlined the origins of the institute in a study for the Australian foreign development agency. His introductory remarks offer a perfect description of its genesis (Sumarto, 2011, p. 2):

The SMERU Research Institute grew out of concerns expressed at the July 1998 Consultative Group for Indonesia (CGI) meeting that there was little independent, reliable, real-time monitoring of the social impact of the economic crisis unfolding in Indonesia. Responding to this concern, a multi-donor initiative led to the creation of SMERU — an acronym that originally stood for "Social Monitoring and Early Response Unit," but now stands alone. The initiative chartered SMERU with a two-year funding package obtained from several donor agencies, including AusAID, ASEM, and USAID, complemented by logistical and administrative support from the World Bank. SMERU's goal was to generate reliable information for the public, policymakers, donors and practitioners on those issues most pertinent to Indonesia to help improve their response to the crisis.

SMERU is, in other words, and by no means is this intended as a critique, a spin-off of a World Bank project that was given a new lease of life by its own researchers. At the time of its foundation, Indonesia was undergoing a dramatic economic crisis and the country and its leaders, as in most crises, were eager to find technocratic solutions to it. SMERU's original approach to "let the data speak for itself" made perfect sense. There was demand for that evidence. Paradoxically, while in Chile it was Pinochet's dictatorial regime that established new processes for technocratic policymaking, in Indonesia it was the emergence of a democratic political system after the fall of Suharto that set forth a new demand for technical expertise and technocratic solutions to the crisis.

SMERU's foundation was a direct response to a demand for evidence-based decision-making. The centre translated this into the undertaking of high-quality analyses, the provision of independent assessments on the effects the economic crisis had had on Indonesia, and the development of decision-making tools for policymakers.

With the crisis over, the institute broadened its scope into a number of areas more suitable for the growing interests of the new government and the international community. Hence, with a grant from the Ford Foundation among others, SMERU embarked on a new phase as an independent research institute.

I spent a couple of weeks at SMERU in April 2011, interviewing its staff as well as other centres in Jakarta. I've continued to engage with the centre and its work over the years. Having worked with SMERU for close to six years before then, I was surprised to find that the concept of a think tank was not very familiar to its staff. Its leadership and researchers had not yet decided how it applied to them. Other centres shared this rather cautionary approach, as we shall see in the case of Peru in the following, and it is far from misguided as some may think. In fact, this caution was well founded based on the institute's origins and the manner in which it developed.

From the start, SMERU focused on the production of evidence and evidence-based decision-making frameworks or tools. In other words, it had both provided evidence and advocated for its use, just as think tanks in the United States have done — as it is clear from the account of the rise of American think tanks in the previous chapter. For instance, one of its flagship projects, a countrywide poverty map first released in 2005, made important contributions to the manner in which the Government of Indonesia undertakes its own analyses. It has been credited with changing both the nature of policy as well as the manner in which policy is done.

By and large, though, their projects have responded to thematic interest and demand from donors and the government, and their approach to communicating their findings has been direct, often in private, and non-confrontational. This technocratic approach to influencing policy reflects the profiles and work practices of SMERU's researchers as well as the institute's own structure.

Rather than focusing on policy issues or themes, or even academic disciplines, SMERU's researchers were in 2011 organised into two main groups according to their research competences: quantitative or qualitative researchers. Much as at the Centre for Global Development in Washington, SMERU's younger staff worked in teams, with junior researchers managed by more than one senior researcher and often moving from one policy issue to another to gain experience and exposure to a broader range of people, methods, and opportunities. This is why bright young Indonesians wanted to work for SMERU. One young researcher there told me that she had joined the organisation because SMERU offered her the opportunity to develop her research skills and gain access to international postgraduate degrees. Unfortunately, her undergraduate course did not offer opportunities to learn and practice research methods and policy analysis.

Donors such as Australian department for Foreign Affairs and Trade have preferred SMERU for similar reasons: Their academic qualifications, which SMERU indicates by the number of papers their researchers publish in international journals, provide them with confidence that the centre's research outputs will not stray too far away from facts. And the government is comfortable in the long-term relationship that has been forged with the institute, partly due to its communication style, which includes movement of staff from SMERU into the public sector.

In a sense, SMERU developed to replace or at least fill the gap left by a number of other institutions in Indonesia. It provides the training or capacity development function that has been overlooked by academia, the independent analysis and technical support that ought to be the role of a more developed and professionalised civil service, and even offers to amplify the voices of the poor and local actors by placing a great deal of emphasis on participatory research methods in their work.

But more importantly, SMERU has created, as in the case of Chile, a home for Indonesian academic researchers to make a living outside of university. The entrepreneurial founders of the centre took on the challenge of transforming a project into an organisation. This is not uncommon of think tanks — and it is certainly the case for many in Africa, South Asia and Latin America. The centres, at their core, serve a rather ordinary purpose: as a source of employment or income for researchers.

To maintain this place in Indonesian policymaking communities and as a go-to think tank for foreign donors, SMERU has invested in its research capacity, possibly at the expense of more proactive and modern means of communication. It has focused its attention on research competencies and skills, developing and maintaining the centre's relationships with key ministries and individuals in the Indonesian public administration, as well as a few strategic donor agencies operating in the country, and carefully engaging with civil society organisations across Indonesia.

When I last worked with SMERU, they had yet to properly engage with the media. Its capacity to help SMERU reach broader audiences had not appealed to the institute's leadership and was in fact been treated with great reservation — overly cautious according to some researchers. This is a challenge that many think tanks face even today, particular if, like SMERU, they emerged from an intensively political environment in which they have developed a strong and successful advisory relationship with their main interlocutors in the government and the local international development community.

This attitude towards the media can be found in other think tanks with similar origins. Researchers and the leaders at Fundación ARU in Bolivia and ZIPAR in Zambia, as well as elsewhere, share this view. However, it is not that they do not recognise the media's importance and the potential benefits for the centres, but that the costs that are associated with a bad experience with the media (namely, reputational costs) are deemed too high.

Engaging with the media, as we shall see, presents a number of fresh challenges that are being ushered by donors' demands of visibility and influence. Unlike the research process, which centres like SMERU know well and can control, the media and its workings are largely unknown to them, or sufficiently unpredictable to be considered as a risk not worth taking.

But the challenge is a pressing one as it refers not just to SMERU's influencing competencies and monitoring and evaluation demands but also to its legitimacy. Nuning Akhmadi at SMERU stated the following:

> *There must be a way that SMERU can be both a trusted friend to Bappanas* (Ministry of National Development and Planning) *and still be accountable to the public.*

SMERU's leadership was well aware of the challenge. How to reconcile its efforts to ensure scientific quality with the media's limited capacity and interest to engage with and communicate the nuances of the scientific method and its often-caveated findings? Unlike think tanks in Chile, born out of political decisions made by an ideologically and politically savvy intellectual community, which included the media and political classes, SMERU is a technocratic construction, designed, and planned to tackle technical problems, including the lack of capacity in these other Indonesian institutions.

Other think tanks in Indonesia have played more partisan and explicitly political roles, more in according with their own origins.

The context in which SMERU developed may further illustrate the challenge at hand for this type of think tank. At the time of its foundation, the Suharto regime had been ruling for 32 years, with long-term negative effects on Indonesia's research capacity. In contrast, in Chile, the new technocratic think tanks that emerged did so at the end of a long period of significant investments in public education and research capacities. The political culture of Indonesian and Chilean researchers could not have been more different.

The intellectuals who set up the Chilean think tanks did so in part to find a platform from where to fight a dictatorship seen by them as devoid of legitimacy. The search for public accountability and acceptance would have been present in the *DNA* of Chilean think tanks, but it would have been unnecessary to an organisation like SMERU, which had been set up at a time when decisive actions and the progressive mantra of evidence-based policy was taking hold in Indonesia and across the world.

However, entirely dependent on their technocratic credentials and without open and public dialogue at the core of their work, organisations like SMERU face the added risk of reputational loss that is less marked for think tanks in more politically open and ideologically diverse environments. Dialogue allowed Chilean and Serbian think tanks to develop and test their theories of the breakdown of democracy and the negative effects of military control over public matters, and their proposals for the future were developed in loose collaboration with others, thus sharing the risk involved in the effort to comment and propose alternative public policy in the context of a dictatorship. In the Chilean case, spreading the risk made sense given that researchers faced more than just challenges to their academic reputation, but this has not been possible in contexts where research funding and work has been driven more by technocratic objectives. In fact, the media had adopted similar strategies that were to continue into its democratic phase in the 1990s (Hojman, 1997).

The accounts of Chilean, Serbian, and Indonesian think tanks suggest that not only does the context matter for the outcome of think tank communities, but that the origin and the qualities and motivations of individuals matter a great deal, too.

The particular case of Indonesia and SMERU is not unique. SMERU provides an illustration of the kind of modern think tank that exists today or is being supported in much of the developing world. In this case, it has a solid research capacity and history, but is now facing an unknown environment. Changes in funding mechanisms and the adoption of new information and communication technologies and approaches are leading to new relationships with donors and clients that prioritise short-term analysis and impact over long-term research investments and relationships, an impersonal communication culture that depends on *sound-bites*, and a changing policy community with new generations of politicians and policymakers, less acquainted with their work and reputation and more with their own advisory networks.

Nonetheless, many other technocratic think tanks are not as well prepared as SMERU is. They are faced with similar challenges but do not benefit from the availability of experienced researchers (as is the case in many African countries, where decades of underinvestment in higher education has reduced the critical mass of researchers) or a rather friendlier funding environment (as is the case in many countries in Latin American or the Western Balkans).

More importantly, though, these challenges affect individuals and not just nameless and faceless entities.

Think Tank Patrons: Everybody Has One

Zhu Xufeng, a Chinese think tank scholar, explained to me that all think tanks in China have a patron within the State and the Party. Independent or private think tanks in particular cannot exist without the patronage of a current or former official. When I share this with my colleagues in the "West," I get the impression that they see this as confirmation that there is no such thing as an independent think tank in China. The concept of independence however needs to be nuanced when we talk about think tanks. As I have considered in the discussion on think tanks' definitions, independence can refer to many things. It is more appropriate to search for intellectual autonomy than some crude definition based on funding, governance, or affiliation. But the Chinese system of patronage does demand that we explore this condition elsewhere.

As in the case of SMERU, the account of the driving forces behind think tanks' origins can offer interesting insights into their development. The cases from Peru and Zambia show how think tanks set up by academics, the corporate sector, donors, and civil society organisations have developed in very different ways, depending on who they owe their dues to.

Academic, corporate, and civil society traditions in Peru: The power of individual researchers

Norma Correa, a researcher at the Pontificia Universidad Católica in Lima, was asked in 2009 to contribute to a scoping study commissioned by the Think Tank Initiative in Latin America. Her analysis conferred particular attention to the origin of think tanks in Latin America.

Of the 18 centres that she considered in her study, 6 had an academic origin and had been founded by private universities; 4 were promoted by the corporate sector, including 2 which were constituted as for-profit firms; 16 were set up as NGOs; only one had been set up and was closely affiliated with a political party; and another one was an independent broad membership-based association.

This initial affiliation has important consequences for their development. To begin with, academic and NGO centres, for different reasons, have deep-rooted participatory structures and processes. Several "democratic" layers govern the selection of their senior leadership, including their directors. Members of an assembly, senior staff, or academic departments, as in the case of academic centres, elect senior management teams or councils and the centres' directors. None, according to Correa's research, appoint their directors from outside the organisation. This practice is extended to their regular staff, with a few appointing researchers via public recruitment processes, employing instead their own academic and professional networks.

Corporate think tanks, particularly those set up as for-profit firms, on the other hand, present more formal and professional recruitment processes, including the search for and formation of management cadres. At the very least, they appoint directors and senior researchers from outside the organisation's immediate networks.

Their origin also affects the manner in which their research agendas and business models have developed. Correa's account shows that corporate centres are more likely to be driven by demand and focus their work on consultancy-style research and short-term to medium-term projects. However, long-term corporate and funder links also create opportunities for corporate support for medium- to long-term policy advocacy.

Academic centres and their researchers are more likely to enjoy a certain degree of freedom to set their own agendas, as they are able to draw resources from their hosts. In exchange, however, universities demand that researchers dedicate an important amount of time towards teaching and other administrative tasks. To complement the basic salaries that they receive from universities, researchers seek consultancies from which the think tanks draw an overhead. Although long-term funding and free funds are rare for academic think tanks, researchers in these centres are able to maintain a long-term line of research, thus creating a demand for it or developing their own expertise on the subject, something that can come in handy when policy windows open up.

Finally, NGO think tanks are more likely to be driven by political discourses. Like other NGOs, they are commonly under pressure to be visible in the eyes of donors and other international agencies. Their funding is linked to their participation in the promotion of distinct policies or ideas, whether they may be ideologically identifiable or not.

In all cases, it is the researchers themselves who drive these agendas and their centres' interactions with their publics and clients. The think tanks often take second stage to their experts, whose publications, media appearances, participation in projects, and secondments to public bodies are kept at arm's length by carefully worded disclaimers.

The role of individuals

In all these cases, as in the case of BCSP in Serbia and think tanks in Chile and the United States, the role that individuals have played in their formation and development is essential to the story of think tanks in Peru. The accounts of the formation of some of the country's most emblematic think tanks, IDL, DESCO, GRADE, Macroconsult, and Apoyo, for example, are closely related to those of their founders and leaders. IDL may not be well known, but its founder, Hernando de Soto, is. DESCO was established in 1965 by a group of professionals and technocrats associated with the Catholic Church and the liberation theology movement. Its first director, Hélan Jaworski, was also one of the founding members of GRADE in 1980. Jaworski, a lawyer, founded GRADE with Francisco Sagasti, an engineer and future President of Peru (2020–2021), and Claudio Herzka, an economist. Herzka was also closely associated with the group of individuals who founded Macroconsult in 1985 after years of working at Peru's Central Bank, where they realised that there was a dearth of sound economic analysis and advice.

Apoyo, one of the organisations that most Peruvians would associate with the idea of a modern think tank, was set up by Felipe Ortiz de Zevallos in 1977, whose role among Peruvian intellectuals is also well known. He has been a principal professor and chancellor at the Universidad del Pacífico, home to another well-known think tank, CIUP, and Peru's ambassador to the United States.

The role of individuals so prominent in the early history of think tanks in the United States is best understood in their influence on the functioning of the centres and on the think tank community as a whole.

There is a clear difference between the more corporate think tanks and academic ones that reflect the role that individual founders have played in the centre's development. In the case of several corporate centres, such as Apoyo, the Instituto Peruano de Economía, and more recently Videnza, the founders have remained at the helm but have created spaces for new generations to join the organisation's senior management positions, thus influencing their development. The same cannot be said about the more academically focused think tanks, particularly those associated, even loosely, with universities such as CIUP at the Universiad del Pacífico, or the Instituto de Estudios Peruanos (IEP), as well as those initially established as NGOs, such as GRADE or DESCO, yet with academic aspirations. These centres' governance, business models, research agenda setting processes, and communications are strongly affected by the prominence of a constant group and type of individuals who have played central deciding roles throughout their history.

Individuals are for the most part responsible for the sustainability of the organisations and the development of their research agendas. Not surprisingly, collective action and cohesive messages are key challenges for these centres. DESCO, owing to its clear ideological origin, was possibly the least fragmented of these examples, but is also facing these challenges. Norma Correa's account of these organisations and my own interviews with staff from CIUP, GRADE, and IEP confirm this.

CIUP, even as an academic research centre, has always strived to be relevant to policy and affect policymaking. Nonetheless, its staff is mainly drawn from within a multi-disciplinary university and is therefore relatively diverse — in interests, ideological preferences, and the public or audiences it targets. While the centre provides a home for researchers and offers them the services they need to develop their academic careers or pursue consultancy opportunities, it does not define *a priori* its research agenda or ideological position. This relative autonomy is also true within thematic programmes. In practice, it means that CIUP does not have a unified and coherent position; rather, its value lies in its experts — what they say is, mostly, up to them.

GRADE, although an independent centre in that it is not associated with any other organisation, formally or informally, faces the same situation. Its senior researchers enjoy a degree of independence that is more expected of tenured academics than of think tank researchers or consultants. They are able to seek and secure contracts on behalf of the think tank and are not expected to seek formal approval from the organisation. The same is true for their engagement with the media or policymakers; each researcher is free to publish and communicate their findings and views independently. It is not surprising then that these sometimes contradict those of other members of the research staff.

This apparent tension is not necessarily a weakness. It allows these think tanks to engage with different audiences even if political regimes or preferences change.

GRADE's senior researchers come together around a brand that enjoys a strong reputation and a business model that facilitates their practice; but they are not directly employed by the organisation. A similar model is found at the IEP. There, senior researchers are not full-time employees of the think tank. Almost all of its assembly members, IEP's senior partners, are based in Peruvian or foreign universities. IEP employs administrative, communication, and research support staff and thus provides a similar role as GRADE to its senior researchers.

Like GRADE and CIUP, IEP has little control over its researchers' policy views or positions in the public sphere. Also like GRADE and CIUP, it has, in the eyes of the public, a recognisable trajectory. While CIUP and GRADE have been associated with a more liberal agenda and economic policy and analysis, IEP is perceived as more progressive and concerned with social policy and qualitative analysis. Within these broad categories, differences are not too difficult to overcome and manage.

Nonetheless, and for different reasons, this model has the same effect as observed in SMERU in relation to the media: an innate institutional distrust. While these think tanks allow their researchers to engage with the media, they do not have a formal strategy that all must follow and instead prefer less visible channels of influence. In a way, each researcher is free to take and face all the risks that this and other new kinds of engagement would bring. The institutions, on the other hand, are sheltered by their stance.

This is all changing, of course; certainly, GRADE and IEP have embarked on important investments and reforms related to their organisations' communications strategies. But it would be naïf to expect a reform that would transform them into politically active think tanks.

In Peru, besides the obvious patronage that exists towards universities, donors, parties, and the market, another kind of relation of dependence exists. Think tanks are, for the most part, dependent on their researchers. But not just as companies need their staff; these think tanks are their staff.

NGOs, the government, foreign donors, the Church, and political parties in Zambia

When I visited Zambia between 2011 and 2012, I explored the roles of four think tanks: the Centre for Trade Policy and Development (CTPD), ZIPAR, the Jesuit Centre for Theological Reflection (JCTR), and then recently formed Policy Monitoring Research Centre (PMRC). Together, they offer another account of think tanks that illustrates how their institutional patronage and affiliations as well as origins can shape their development. These cases also show the variability that is possible even within a single country, thus making it difficult to talk of a particular national think tank tradition.

Unlike the Chilean story where think tanks emerged out of a highly ideological political struggle or the entrepreneurial nature of a small group of technically driven researchers and policy entrepreneurs in Indonesia and Peru, African think tanks appear, at least in some cases, to be more the expected outputs of donor interventions (Kimenyi and Datta, 2011) or the agency of international development actors. The first three examples illustrate how this external patronage can affect the formation of think tanks but also how difficult it is to make any generalisations across regions or even within countries. They also usher in the role of an unexpected, but by now familiar, player: the church. The final example presents a welcome alternative for African think tanks: political or partisan patronage.

An NGO network

I first came across CTPD through its former director, Saviour Mwambwa, whom I had met in a previous position at a Zambian network of civil society organisations with which we had collaborated. When I visited Lusaka in May 2011, I was introduced to the centre as a network of civil society organisations working together to promote more equitable trade policies. Its members are local and international NGOs operating in Zambia that, as in many other countries, had decided to set up a network to coordinate their actions on a particular policy issue — trade, in this case. The secretariat, CTPD, was modestly staffed and funded and had been largely responsible for the implementation of projects designed by their members and other funders in Zambia, many of which involve capacity-building and service provision to its members and their own beneficiaries.

When it was first set up, and as late as 2010, CTPD enjoyed core or long-term funding from some of its international NGO members. However, as funding trends changed for these organisations, they too enforced new contractual arrangements on CTPD that had brought it to the brink of financial insolvency. Unable to carry funds from one financial year to the other or to use these to cover the organisation's overhead, it could only operate if it was able to secure new contracts before the end of any financial year.

As a network, a board that represents the membership of local and international NGOs governed the centre. Most of its funding and activities at the time came from a few international NGOs in the form of short-term projects. It is not surprising then that the concept of a think tank seemed somewhat unrelated to CTPD. When I asked, neither its leadership nor staff immediately recognised it as such. At an event with other NGOs that CTPD organised, I asked the participants about think tanks and, again, nobody considered it to be one. Instead, participants described think tanks as being "of the government," "focused on economic policy," "neo-liberal," or "academic."

Slowly, and largely led by its director, the organisation began to emerge from the comfort of the civil society sector from which it drew its funding and networks to engage more directly with policymakers and international funders on a broader set of policy issues including tax (particularly that related to the mining sector), international trade, and investment. This move from secretariat to think tank paid off. CTPD was able to make a name for itself among economic policy and research funders in Zambia by means of a few policy-focused studies and outputs.

The challenge that CTPD then faced was how to maintain its links to civil society organisations while it left behind the constraints that they and the network governance imposed. CTPD, after all, drew its legitimacy from its capacity to represent, engage, and involve civil society organisations working across Zambia. Its civil society patrons may have been keeping it from moving forward faster, but they also provided the centre with a platform from which it could work.

CTPD's trajectory was not unlike SMERU's — even if the organisations are remarkably different. Both have emerged as spin-offs of temporary projects or initiatives. SMERU was a World Bank project designed to respond to a particular national context in Indonesia, while CTPD was conceived as a network to address a pressing issue in Zambia. The agency of its staff and particularly their leaders has driven them in a similar direction.

A project of the Ministry of Finance

In August 2011, I was asked by the U.K. Department for International Development (DFID) in Zambia to undertake an organisational assessment of ZIPAR. DFID (now FCDO) had been considering how to best support the think tank and expected my analysis to provide it with some guidance. I had visited ZIPAR before in May while working with CTPD, and so was eager to follow up on what I had already found about the centre.

The ZIPAR model offers what feels like a more typical African think tank story and illustrates what CTPD and its peers perceived a think tank to be. It also represents the extent of most African governments' embrace of the idea of evidence-informed policymaking, much like what SMERU found in Indonesia. The consensus among donors, researchers, and policymakers is that economic research capacity in Zambia is limited. So ZIPAR was originally set up as a project of the Ministry of Finance to fill that gap. It is, however, certainly not just another government think tank.

While the Ministry of Finance provided ZIPAR with its offices, funded its administrative staff, and had an important role in its governance, chairing the think tank's board, for example, other funders were involved in its setup. Its senior research fellows were supported by a grant from the Africa Capacity Building Foundation (ACBF), mainly directed towards academic research and publications. The ACBF has played a leading role from the start, as it has in other African countries where it has also supported the formation of similar government-linked economic policy think tanks. There were also a couple of projects indirectly funded by the Danish cooperation, which provided ZIPAR with the necessary funds to employ younger researchers-cum-consultants.

Finally, DFID also came on board with economic research technical assistance, in the form of 50% of the time of its own economic adviser and funding for additional research costs as well as communication activities.

Not surprisingly, each funder had different expectations of what ZIPAR's role should have been. The Ministry of Finance and other government departments expected ZIPAR to undertake policy-relevant research and offer private, direct, and impartial advice to the government. Staff at the Bank of Zambia had very high expectations for ZIPAR and its contribution to policymaking in Zambia — helping to anchor economic policymaking to robust evidence. None, however, expected ZIPAR to drive the public debate or attempt to shape and set the agenda. They saw ZIPAR very much as providing evidence to legitimise (and operationalise) government policy — very much so as the primary function played by think tanks in Southeast and East Asia (Nachiappan *et al.*, 2010) or contracting think tanks around the world more generally. This expectation, however, has a different origin. The language of results in international development, alignment to government policies, and demand-led research all conspire to implant, in the minds of policymakers, the idea that research, just as other development intervention, must serve the agency of the government. Only when a demand is fulfilled is it equated with impact and value for money.

But whereas the government would like this impact to take place in private, through presentations or participating in executive or parliamentary committees, its other funders expected a more public role for the think tank. As many donors today, DFID at the time was very concerned with the visibility of its development funding, and so would prefer a more public and open approach to policy influencing. This demanded shorter studies, the analysis of current issues, the publication of opinions, working with the media, organising public events and debates, making use of new communication technologies, etc. None are strategies easily pursued by a think tank hosted by the Ministry of Finance.

ACBF's academic concerns added a dollop of complexity to the mix. In its first full year of operation, 2011, ZIPAR focused its research not on short- to medium-term studies and the active dissemination of expert opinions or policy proposals but rather on a fairly large and academic research programme. This is not surprising: the researchers had academic backgrounds and were hired for those credentials, their research projects involved primary data collection, and they had been actively discouraged from publishing opinions and analyses by a quality control process that was more apt for an academic research centre than a think tank.

This interesting and possibly contradicting combination of supporters and demands illustrates the complex nature of think tanks in many African countries, which owe their existence to multiple uncoordinated patrons. What then is ZIPAR? Or better yet, what does ZIPAR consider its role to be, in light of this complexity? Its first director, its staff, funders, and main audiences described it in several ways: as a semi-autonomous or semi-independent research centre, a quasi-governmental body, a project of the Ministry of Finance, a civil society organisation, and a not-for-profit consultancy. Clearly, this depends on who one asks.

One thing that set it apart from other think tanks is its genesis, even if it shares some similarities with its Zambian peers. Unlike SMERU, which was set up by the agency of its staff, Peruvian think tanks that were mostly founded by entrepreneurial technocrats and academics, Chilean think tanks that were set up academics and public intellectuals, and CTPD which was being transformed by its own director, ZIPAR, as it stood in those early years, was conceived as a project by its patrons. Its organisational structure and core administrative functions were established by the centres' funders before any researchers had been hired and any research had been undertaken. Without doubting the commitment of its staff, ZIPAR struggled to find an internal champion: Its board was composed of representatives of other institutions (the government, donors, the private sector, academia, and civil society), and its leadership and staff were still only employees.

This approach is reminiscent of the initial separation of funder, citizen, and expert that the large professional philanthropic powerhouses experienced in the United States in the early 1900s. This very same approach is observed in much of the funding available for other think tanks. Interestingly, however, the same forces affected CTPD and ZIPAR, but these organisations moved in two entirely different directions, and this is in part down to the role that individuals have played and their positioning in relation to government.

The dominance of academic research as the source of credibility and legitimacy also sets ZIPAR apart from organisations like CTPD. For CTPD, the absence of senior researchers among its staff meant that its work was less likely to be considered as sufficiently technical by the economic research and policy communities. It must therefore compensate for this by developing high-profile initiatives (such as a Tax Justice Platform), new communication outputs and channels (such as policy briefs and local and international media appearances), and maintaining close ties to the NGO community in Zambia from which it draws its legitimacy.

A holy patron: The Church

Despite their uniqueness, it is not difficult to identify ZIPAR and CTPD as rather "typical" think tanks: Semi-autonomous public think tanks are common in Europe, and CTPD's network governance is not too dissimilar to that of many Latin American NGO think tanks. They draw their legitimacy from a combination of sources, but research-based evidence plays an important role — more so in the case of ZIPAR. Both certainly aimed to influence policy in Zambia and fulfil, by design or otherwise, several think tanks' functions.

They were not, however, entirely autonomous. ZIPAR's closeness to the government and its dependence on ACBF's design and DFID funding directed both its research agenda and how it went about implementing and communicating it. Limited by the interest of its board and the short-term project funding available to it, CTPD was in a similar position.

Another interesting think tank in Zambia is the Jesuit Centre for Theological Reflection (JCTR). It merits a mention for being, at least at first sight, a rather unique centre in that it presents itself as a research organisation that combines research-based evidence with appeals to action based on "social teachings" drawn directly from Biblical scripture. As part of the Church in Zambia and Malawi, JCTR was able to draw from its own resources, both in kind and in cash, to reach an already interested audience that is naturally prone to pay attention and award it the legitimacy that other think tanks have to earn through research, outreach to civil society, and new affiliations. This, ironically for an organisation that exists within the Church, gave it a sense of autonomy to pursue the arguments it wished to that neither ZIPAR nor CTPD enjoyed.

The Chilean case awarded the Catholic Church a critical role right at the start of the emergence of modern think tanks in the early 1970s. In Peru as well, proponents of the liberation theology movement with links to the Catholic Church in Peru set up DESCO. The Universidad del Pacífico, which hosts CIUP, is a Jesuit University. Rather than this being an argument in favour of more religious think tanks, I think it illustrates the importance that narratives and ideology play in galvanising the efforts required in the formation, development, and funding of think tanks.

In this case, the JCTR was rather effective in using its patron to achieve its objectives — an example that illustrates the potential benefits that patronage offers to think tanks.

One of the JCTR's main initiatives was the Basic Needs Basket. This is a project that involved monthly measurements of the cost of basic needs such as food, transport, and housing in Lusaka, 12 other towns, and 5 rural areas. The findings were published in a simple and straightforward one-page document and disseminated through JCTR's and the Church's own network of employees and volunteers across the country. Frequent analyses were also undertaken and published in pamphlets and other media.

The effect was remarkable. After a few conversations with taxi drivers in Lusaka in May and August 2011 about the upcoming presidential elections, I soon realised that many were referring to the same figures when discussing the cost of living in the city and their demands from the presidential candidates. It was not difficult to track these figures back to JCTR's Basic Needs Basket study, publications, and outreach efforts.

Evidence, the facts alone, cannot explain this success. Evidence combined with explicit appeals to values, and the stories drawn from the Bible more specifically, makes JCTR an interesting case of an ideologically identifiable think tank (but not one that Josef Braml had in mind when he compared think tanks in the United States and Germany). JCTR was able to use the well-known and powerful narrative, as well as the many stories and metaphors, that religion provides to communicate its arguments in a manner that resonated with both specialised and general audiences in Zambia. It is not surprising then that JCTR made more and better use of the mainstream media as well as social networks (physical, not virtual) to disseminate its messages, for instance, with op-eds in a national newspaper, *The Post*. And this was probably influenced by the staff that it was able to attract: interested in the opportunities for research and influence that the centre presented, more than by its religious stance.

In contrast, this appealing and attractive narrative is something that ZIPAR and CTPD had to develop themselves. Driven by technocratic imperatives (the evidence-informed policy mantra that has been adopted by the international development community) and short-term contracts to deliver projects devised by its funders, ZIPAR and CTPD, respectively, enjoy very little freedom to develop cohesive or complete narratives of their own. JCTR had a ready-made one.

Evidence plays different roles in these organisations, but it is no less important for them. In JCTR's case, however, evidence is used in very much the same way as it is employed in other ideologically identifiable think tanks. The Heritage Foundation in the United States and the Fabian Society in the United Kingdom work in a similar manner to JCTR: They use evidence to strengthen their narrative: conservative and progressive, respectively.

It is then not only the Church that brought the stories of think tanks in Chile during the 1970s and JCTR together. Instead, it is the presence of a broader political narrative; in response to the liberal policies and authoritarian politics of the Pinochet regime, and the Christian message of social justice. These narratives make it easier to attract support and interest from their funders and publics, and at the same time nurture a cohesive and consistent organisation, staff, and message.

It is important to note, before bringing this section to a close, that the emphasis I have given religion, particularly in Africa, is a perfect illustration of how perspective affects the study of think tanks. For an atheist

Latin American whose career started and developed in the United Kingdom, the positive role that religion plays in these cases comes as a surprise. Even in the U.K. (where the head of State is also the head of the church) or in Peru (where the Catholic Church plays an important public and political role), an organisation like JCTR would seem out of place. However, what JCTR represents and proposes is by no means out of place in Zambia and many parts of Africa where religion plays a central role both in private and public life. And any attempt to study think tanks in greater detail in these circumstances should take these matters seriously (Broadbent, 2012).

Political parties as patrons

A recent development in the Zambian political scene was the foundation of the Policy Monitoring and Research Centre (PMRC). As its name suggests, it was set up to monitor the new government's plans as described by the Patriotic Front's manifesto. What sets it apart from other think tanks that adopt monitoring functions is that PMRC was established by the chair of the Patriotic Front to keep his own party on track and focused on delivering on the promises it made during the 2011 presidential campaign.

PMRC is not formally affiliated with the Patriotic Front, but, like its counterparts in Europe and the United States, close ties with the party's leadership allowed it to gain information, access to policymaking spaces, and offer ideas and advice. Unlike ZIPAR and CTPD, two think tanks that are well within their own sphere of operations, PMRC has been set up outside the aid sector.[2] Its main promoters, its director and staff, had only limited experience working with aid agencies and engaging with the development policy discourse. Instead, their networks were dominated by Zambian politics and the private sector, and their inspiration for PMRC

---

[2] The British Department for International Development funded PMRC's start-up phase on request from the party's leadership. It saw it as an opportunity to develop new institutions to support the programmatic capacity of Zambia parties as well as a symbolic gesture to the new government.

This constituted a bold and risky move for DFID. Unfortunately, it has not been followed through. Nonetheless, the foundations left, as well as the experience, are of great importance for the future of Zambia's think tank community.

owes more to the kind of corporate think tanks found in the United States or the United Kingdom.

PMRC's setup and organisational culture were also very different from those found in ZIPAR or CTPD. Instead of academia or NGOs, PMRC's founding director's background was in the private sector. Her approach was more corporate and driven by results than those found in other organisations in the sector, where academic imperatives or dialogue and participation were given priority.

This difference in approach has translated into an organisation that more closely resembled London-based political think tanks than anything found in Africa at the time. The organisation was politically minded, its research was informed and guided by political and policy questions, communication was incorporated into their efforts right from the start, and it deployed a comprehensive communications strategy to accompany and support its research and political engagement.

PMRC has continued to evolve since my last engagement with it. Its leadership and patronage changed — from one political party to another — but many of its functions have remained, and its political relevance has remained.

In a way, this new member of the Zambian think tank family helps us to understand the older members. PMRC is how it is because of why it was set up and by whom. ZIPAR's academic style, CTPD's participatory and capacity-building focus, and JCTR's value-based approach can also be explained by their origins.

Emerging Ideas

From the cases presented here, we can identify some emerging ideas. Some will be further explored in the following chapters. Whereas the history of think tanks in the United States presents a consistent story — the search for (or the illusion of) pragmatic evidence-informed policymaking — these offer a more real, complex, and "messier" account of think tanks across the world.

It appears, for instance, that think tanks have thrived even during periods of authoritarian regimes. In Chile, and against all odds, think tanks carved out spaces for themselves. In Serbia, too, think tanks emerged as part of the democratic struggle. Democracy and political liberalisation, *contrary to* what is often reported, do not necessarily seem to be a driver of think tank formation; in fact, their absence can be a cause of think tank activity.

In Latin America, and in other regions, the Church has played a critical role. It has provided a useful space for the development of think tanks and continues to support their formation and agency. This makes sense. We tend to associate international development with secular interventions, but in fact much of development assistance is provided by faith-based organisations. And before these existed, the Church provided the networks through which ideas and resources flowed from developed to developing countries. Before the United Nations, DFID, or the World Bank deployed development consultants, priests and missionaries from Europe and the United States made their way to developing countries. A great deal of this support was in the form of education, and this led to the formation of research centres and, today, think tanks.

The role of individuals is also particularly interesting. In the cases described from Peru, individual researchers have played critical roles in setting up think tanks and in their governance structures and practices. They remain, in some cases, more powerful than the centres themselves, and this has important consequences on the think tanks' capacity to set and develop their research agendas, mobilise resources, engage with others, develop and promote cohesive and sustained policy arguments, communicate them, etc. This, however, is by no means the destiny of all such think tanks. BCSP's founding leaders chose an alternative path and, in taking a step to the side, freed their organisations to forge a new identity.

Think tanks' institutional associations (for instance, with political parties and politicians, the corporate sector, the State, civil society organisations, and academia) also play a role in their development. Their business models, the manner in which the research agenda is developed, the choice of policy audiences, etc., are clearly influenced by these. And so are the functions they are able to fulfil. More academic think tanks, or those that are able to pursue more independent research, tend to shun influencing, legitimising, and auditing functions; instead, they prefer to adopt longer-term and more indirect ones, such as developing future generations of researchers and policymakers or educating the public and the elites. Corporate associations, as well as close links to the State, on the other hand, seem more closely linked to those more pragmatic or technocratic functions, evolving in functions and structure as necessary.

Inevitably, in the absence of sufficient domestic funding, think tanks in the Global South owe much of their existence to foreign funding and support. There are, however, differences in the way in which this support has been used by their beneficiaries. Whereas SMERU's researchers have taken on the initiative, it is yet to be seen if organisations like ZIPAR, funded by a coalition of actors that include the government, will be able to. While its staff in effect designed SMERU, ZIPAR still remains a development project subject to embedded donor presence. Hence, the balance between the agency of the funders and the think tank emerges as an important element to consider. Who drives whom?

The internal governance of think tanks also appears to be of importance. This can be explained by paying closer attention to the context in which think tanks have been set up. In Peru, the prevailing model among the think tanks not linked to the corporate sector is one where individuals have come together in the form of formal and informal associations. They interact with each other following "democratic" principles and processes that affect the centres' governance: for instance, limiting the appointment of directors and senior managers from within the organisation. This reflects a model more closely related to the original think tanks of the late 1800s in the United States, but more importantly, a long historical tradition of academic and intellectual associations that goes as far back as the late 1700s in Peru.

The role of funding, but particularly the type of funding, is also of great importance to think tanks. This has an effect on their own capabilities in setting their research agendas and choosing their influencing strategies. Funding types, the role of individuals, and the influence of foreign donors appear to be conspiring towards reducing the space for manoeuvre of think tanks.

Overall, the importance of perceiving think tanks as a fluid concept and recognising that the organisations that adopt the label are inevitably subject to changes emerges as a crucial idea. Already, this forces us to emphasise the importance of adopting flexible boundaries in the definition of think tanks and considering the effect that think tanks' contexts, namely, their politics, have on those boundaries over time.

Unfortunately, these accounts, with the possible exception of the Chilean case and PMRC's recent appearance in the Zambian scene, remain relatively neutral to the issue of partisan politics. Some hints as to its importance in other cases can be inferred, however. SMERU emerged out of the opportunities and challenges that the transition to democratic rule presented to Indonesia. In Peru, think tanks have in general maintained some distance from their researchers' views to avoid political backlash. And in Zambia, ZIPAR faces a difficult political dilemma, fuelled in part by its funders' own conflicting interpretations of the role of think tanks in society.

Finally, these cases demonstrate that the organisational perspective is useful but not sufficient for the study of think tanks. It does not adequately provide an explanation for the nature of the organisation being studied.

The next chapter will address the issue of politics and political institutions more directly as a way of introducing the other perspectives considered in this book.

Chapter 6

Political Lenses

The history of think tanks in the United States and the cases from across the rest of the world presented in earlier chapters can help us understand the complex emergence and diverse nature of think tanks. It should be possible by now to recognise the multiple powerful ideas that have driven their formation; the different types of expertise that they have prioritised; their different business models; the roles of individuals versus institutions; and the changing nature of think tanks' engagement with each other, their funders, and the public more generally. However, so far, the organisational perspective cannot account for all these nuances, and hence the picture that emerges is a somewhat messy one. We need an alternative perspective or narrative to make sense of it all.

In addition, the study of think tanks in the Global South or in contexts where institutions are weaker or less developed demands that we pay greater attention to the differences and similarities that exist between them, and to do so we must focus our attention beyond the organisations themselves and instead look around them: at their contexts. As the cases presented earlier demonstrate, it would be irresponsible, and futile, to attempt to compare them like for like without taking their distinct political contexts into account — contexts that explain both their origins and subsequent development.

This section provides a brief outline of different analytical frameworks used to study and conceptualise the political contexts of think tanks in Latin America, Africa, South Asia, and East and Southeast Asia. These studies offer interesting insights into the nature of think tanks in different regions and, I hope, shed some light into the complex and incomplete stories presented previously. They illustrate the way in which elitist, statist, and pluralist perspectives, which will be described in further detail later on, can help to study think tanks.

Unlike the history of think tanks in the United States, I do not set out to offer a full account of their development (nor the most up-to-date literature); instead, I only offer a glimpse into the critical influence of politics on think tanks across different contexts.

In Latin America, the emphasis is on the degree of institutionalisation of the political space (understood as both the development of the institutions of the State as well as political parties and other policy actors such as academia and the media) (Mendizabal and Sample, 2009; Correa and Mendizabal, 2011) and the roles that political and technical elites have played in certain cases as opposed to the more pluralistic political contexts in others. In Southeast and East Asia, the focus is on the interplay between the politics of production and the politics of power found in the different models of the Asian development state (Nachiappan *et al.*, 2010). The State (whether in the form of single-party rule or corporatist) plays an undeniable leading role. In South Asia, the analysis pays special attention to the balance between the State's exercise of despotic and infrastructural power (Srivastava, 2012) and the role that the civil service has played in opening and closing spaces in which think tanks can operate. Finally, in Sub-Saharan Africa, two key dimensions of concern emerge in the study of think tanks: the politics of State power and the politics of external influence — and how they have competed with each other (Kimenyi and Datta, 2011; Da Costa, 2011). The case of Africa also offers an interesting opportunity to assess the role of foreign donors.

Latin America: Searching for the Goldilocks Zone

Inspired by earlier research by Adolfo Garcé on think tanks as well as the work promoted by the International Institute for Democracy and Electoral Assistance (IDEA) on political parties in Latin America, Kristen Sample, IDEA's Director for the Andean region, and I set out to explore the roles that think tanks played in Latin American politics by focusing on their relationship with political parties (Mendizabal and Sample, 2009). Adolfo Garcé (2009, 2018) developed an analytical framework that inspired and informed a series of cases from Bolivia (Toranzo, 2009), Colombia (Londoño, 2009), Ecuador (Bellettini and Carrión, 2009), Peru (Tanner, 2002), and

Table 1. Think tanks and political parties in Latin America.

Weak political parties and poorly connected to think tanks	Weak political parties but good links to think tanks
Strong political parties but poorly connected to think tanks	Strong political parties and good links to think tanks

Chile (Cociña and Toro, 2009). Later, with Norma Correa, we expanded the analysis to other players, including policymakers (and bureaucracies) and the media (Correa and Mendizabal, 2011).

Garcé identified two key dimensions of analysis: the degree of institutionalisation of political parties and their degree of connectedness with external[1] think tanks in their own countries (Table 1).

From his own analysis, he found that the first case (weak parties and poor connections) is infrequent in Latin America. He suggested the case of Bolivia's Movimiento al Socialismo (MAS) as a possible exception, but in a subsequent study, Rafael Loayza Bueno (Loayza, 2011) showed that, in fact, MAS had developed strong ties with several Bolivian think tanks, which were instrumental in developing some of Evo Morales's key policies.

More common are the cases of weak parties with close ties to think tanks (for instance, in Ecuador and Peru) or strong parties with poor links to think tanks (as in Uruguay). Less frequent at the time was the Chilean case where strong political parties have enjoyed good relations with external think tanks — something that is increasingly found in Brazil and Argentina and, historically, in Colombia.

In Argentina, Fundación Pensar and the PRO party enjoyed a close and supportive relationship (Echt, 2019).

The same idea, that is, the more developed and well defined the parties, the stronger and more productive their ties with think tanks, was later applied to government bureaucracy (Tanaka *et al.*, 2011) and to the media (Uceda, 2011; Livszyc and Romé, 2011). In these cases, too, the degree of institutional maturity of these other policy players can help explain the strength of the think tanks themselves. Programmatic parties, for example, are able to demand more research and advice and so provide think tanks

---

[1]He sought to differentiate between external think tanks and think tanks set up by and operating within political parties.

with formal and reliable communications channels that offer useful opportunities for intellectual feedback. An intelligent buyer, so to speak, also provides incentives for think tanks to invest in their own capacities.

Conversely, these other institutions will be less likely to demand and use research and advice if they are themselves not developed the necessary competencies. Consequently, it is also less likely that think tanks will want to engage with them if they are not developed enough to guarantee a certain degree of stability and constructive interaction. Fearful that the relationship will turn sour in cases of differences of opinion or that their reputation will suffer by association, think tanks and individual researchers will be wary of establishing formal links with parties or political players. They will also not see the necessity of investing in their own capacity if their prospective collaborator is poorly informed, incompetent or, even, corrupt. Think tanks' capacity is hence defined by others around them (Medvetz, 2008).

The cases explored in the studies of think tanks and political parties (Mendizabal and Sample, 2009) and think tanks and policymakers and the media (Correa and Mendizabal, 2011). Their beginnings can be traced as far back as the late 1700s when Peruvian intellectuals formed an academic society called La Sociedad de Amantes del País in 1790 to discuss and publish their views on the current political, social, and economic issues and broad recommendations for an independent Peru (which would not happen for another half a century). In Colombia, the early origins of think tanks can be found in the periodical publications that set out the initial terms of the intellectual and political struggle involved in developing the republic in the second half of the 19th century (Londoño, 2009). Later on, think tanks developed faster, as in the case of Colombia, where political parties and the political system developed, too.

The sudden growth of think tanks observed from the 1970s onwards, though, is undeniably down to the funding provided by international agencies who, for different reasons, sought to support applied research in the region. In some cases, as in Chile, this support fell on fertile ground, with strong political, private, and civil society institutions.

So, is good governance — such as having strong and mature party systems and civil services — a condition for the development of think tanks? The Latin American experience offers an inconclusive answer to this question.

It is certainly true that think tanks are more common, better developed, and enjoy more constructive relations with other political players in the more developed political systems in the region. Chile is a clear example of this. However, many of Chile's think tanks were in fact founded and thrived during a period of authoritarian rule. The political coalition that ruled Chile since the fall of the Pinochet regime for over two decades until 2011, according to some, contributed to the development of a less dynamic think tank community. Fernando Londoño reported the same in the case of Colombia. He found that long periods of political consensus led not to more but less think tank activity.

In other words, to thrive, think tanks in Latin America have depended on a certain degree of political polarisation or competition — not too hot to bring down the political system and think tanks with it, and not too cold to reduce ideological and technical debate: a sort of Goldilocks Zone for think tanks.

Current concerns about political polarisation (On Think Tanks, 2024) contrast with these findings.

What made a difference was that the relations between think tanks and parties in Chile were strong. Both the right and the left developed strong ties with researchers and experts so that even in an undemocratic context, think tanks were able to develop. Political competition in Chile and Colombia led to more demand for evidence from think tanks and to the strengthening of the ties between politicians and technocrats. When Chile's and Colombia's coalition governments took power and the competition between parties fell, so did the demand for new ideas, and the ties between think tanks and parties were weakened.

In Peru or Ecuador, political competition may be much higher (in 2025, more than 40 parties are planning to contest the 2026 presidential elections). However, institutions are weak and so are their links to think tanks and the think tanks themselves.

Other trends are discernible, too. Peru offers an illustration of how different political contexts have led, as expected, to the foundation and formation of different types of think tanks. In the 1980s, after more than a decade of military rule, political parties and think tanks thrived, particularly those associated with left-wing ideologies. The 1990s, on the back of the country's worst economic crisis in recent history, ushered in the collapse of the political system and the introduction of structural adjustment measures that saw the rise of liberal and technocratic think tanks that replaced, de facto, any lingering planning responsibilities still held by the public sector and political parties' programmatic capacity. The 2000s, a period of significant consensus, did not see the formation of many new think tanks; this changed towards the end of the second decade of the 21st century with, for example, Equilibrium CenDe (founded by Venezuelans in response to the migration crisis), Urbes Lab (founded by recent graduates in response to a deepening crisis in urban centres), Red de Desarrollo (REDES, founded by the private sector in response to worsening public discourse), and the Centro para el análisis de políticas públicas de educación superior (CAPPES, in response to the unravelling of the education sector reforms of the previous decade).

Paradoxically, as in most Latin American countries, economic liberalisation reduced rather than increased the intellectual space, which was taken over by technocrats deployed by increasingly autocratic regimes (Mendizabal, 2022). The reforms demanded by the international community in the 1980s and 1990s were so unpopular that they led to governments reducing the political and media space available for think tanks to explore ideological alternatives. In fact, many think tanks, such as GRADE, CIUP, and IDL, played a part in steering the narrative away from the political and into the safer territory of technocratic terms. The Southeast and East Asian case described in the following provides additional proof that think tanks are not necessarily a feature of open democratic politics.

Of particular importance, and as illustrated by the Chilean case, is that both democratic and military or authoritarian regimes have played critical roles in the opening and closing of the policy space — the latter, however, explaining the introduction and support of technocratic approaches to policymaking and defining the terms of engagement for think tanks and society at large.

The new millennium finally did see the political liberalisation that had been put on hold in many countries during the 1990s. This triggered the development of a new phase of political party politics and a more dynamic think tank community, often (albeit driven by mostly individual researchers) connected to political players through informal rather than formal channels (Tanaka *et al.*, 2009). This fits well with Orazio Bellettini's view that democratisation in the region has led to the formation of more think tanks (Belletini, 2007), although, as we have seen already, the story is by no means a causal relation, and this did not automatically translate into greater plurality.

By incorporating a historical lens, the researchers in these studies were able to draw direct and clear links between the development of other political institutions and the think tanks themselves, thus confirming Medvetz's (2008) reflections that think tanks are better defined by their relation to others. More importantly, though, the appearance and development of think tanks in Latin America, although greatly dependent on foreign funding, particularly since the 1970s, are intrinsically linked to the development of each country's own political systems. In this context, think tanks have been able to develop a broad range of functions — often adapting and changing them over time. Because of this, it is difficult to talk of a single Latin American tradition of think tanks. To truly understand them, one needs to focus on the national level and, even there, take into account the dramatic differences that academic, corporatist, civil society, and political think tanks exhibit.

It is easy to suggest think tanks exogenous to Latin America. I argue that they are an indigenous phenomena. The intellectuals and *libertadores* who first forged the idea and then the reality of independence were influenced by North American and European ideas and support over 200 years ago; but they were "*Americanos*". Modern Latin American think tanks have greatly benefited from foreign ideas and funding; but they are local.

Southeast and East Asia: Like Clockwork?

For Southeast and East Asia, Karthick Nachiappan, Ajoy Datta, and I proposed a different approach to the study of the political context in which think tanks had developed since the Second World War (Nachiappan *et al.*, 2010). Rather than focusing on their relationship with political parties, we took the developmental state as the unit of analysis. Assuming that think tanks were a political actor, we considered that the same factors explaining the state's differences across the region could explain the differences found in think tanks. Nachiappan first introduced the idea of a two-dimensional framework to study the politics of Southeast and East Asia that considered the politics of production, on the one hand, which explained different ways of organising the means of production (market-oriented versus state-led, for example), and the politics of power, on the other hand (concentrated on single parties or leaders or dispersed among more actors in the public and private sectors).

The analysis identified three underlining forces that explain the particular characteristics of the various developmental states as well as the different incarnations of think tanks in the region: the role of nationalism which drove the efforts for state-building in the 1950s and 1960s, the links between the bureaucracy and the private sector, and the concentration of power around individuals, the state, or the private sector.

Unlike in Latin America, we found that think tanks there had not just accidentally emerged out of the interaction of these various political players (academics, intellectuals, politicians, journalists, activists, funders, etc.) but rather had been planned to serve very specific roles assigned to them by the developmental states in which they exist. In general, think tanks in the region, despite their differences, also seem to share a common legitimising function — whether it is to the State, civil society (or particular individuals or economic or political interests), or the private sector.

Beyond this function, think tanks were conceived as participants of national and regional developmental visions, and so it is possible to recognise some trends and traditions. Technocratic economic think tanks emerged in all countries to contribute to the process of economic development, characterised by the pursuit of economic growth (albeit with different approaches to the means of achieving it) as a means of competing with their neighbours. Foreign policy and security think tanks were set up as windows into the politics and strategies of their neighbours and nations further afield. Regional integration think tanks appeared to pursue an agenda that sought to simultaneously increase collaboration between countries while safeguarding their own economic growth and development.

But this is as far as similarities go. Just as in Latin America, think tanks reflect indigenous political expressions, and their differences are more appropriately explained by the differences in the political, social, and economic systems in each country. Think tanks in Japan and South Korea, for example, have closer ties to the private sector; in Vietnam and China, they exist mainly within the policymaking apparatus of the State and the Communist Party's patronage; and in Indonesia and Malaysia, think tanks are more likely to be associated to individuals or interest groups from domestic and international political, economic, or civil society (for instance, religious) communities.

This expression of power and patronage puts into question the idea of independence that is often used in think tank definitions. In East and Southeast Asia, the possibility of complete independence appears impossible to comprehend unless one accepts that although think tanks are by definition dependent on a particular institution (in the bureaucracy–private sector continuum), they can retain certain operational autonomy — even if this has been designed into their set up by their patrons.

Also, as in Latin America, we found an inter-temporal dimension to the evolution of think tanks in the region. In general, the trend appeared to be that in all cases, think tanks were first more commonly set up within the state or close to the bureaucracy but progressively moved away from it, to a greater degree in the more politically and economically liberal nations, as political and economic reforms were introduced. Here, too, democratic periods (such as in Indonesia after Suharto) or economic and political reform led to the formation of new think tanks. In China, in particular, recent economic and political reforms have been accompanied by a significant increase in the number of think tanks (Naughton, 2002; Zufeng, 2009) that, while may appear to be challenging the State, are more likely to be part of the developmental path chosen by it.

However, the causal relation between democracy or political openness and the existence of think tanks can once more be questioned. Indonesian think tanks are faced with a barrage of legislation that greatly limits their capacity to function. From interviews with staff of local think tanks, I was able to identify a number of ways in which government policy curbs think tanks' work: labour legislation that reduced think tanks' flexibility and capacity to hire and retain the right staff, procurement policies that prevent think tanks and other not-for-profits from bidding for public funds, tax law that discourages businesses and individuals from donating to research centres, etc. Furthermore, as we will see in Chapter 7, it was the Chinese State that was responsible for encouraging the formation and development of think tanks in China, and to a lesser extent, the same has happened in Vietnam. Democratic Japan does not have the thriving think tank community, independent of any private or public ties, which one would expect to find in western democracies; however, it has its share of corporate think tanks that reflect the role the private sector plays in its politics.

South Asia: Old (and New?) Nationalism

Think tanks are by no means uncommon in South Asia, and, just as in Southeast and East Asia, domestically funded foreign policy and security think tanks are an interesting feature of the regional think tank community. When the Think Tank Initiative was launched in South Asia in 2010, several Indian commentators and intellectuals triggered a public discussion on the roles of think tanks in India. In August 2010, Sanjaya Baru, a former media advisor to the prime minister, wrote an article for IMAGINDIA arguing that India's best-known think tanks "on economic policy, national security and foreign affairs, were finding it easier to raise funds abroad than at home, be it from a bureaucratic and feudal governmental system or from a miserly and disinterested corporate sector" (Baru, 2010). He argued that India's own philanthropists were supporting foreign think tanks and research centres such as Yale or Carnegie but not their own. As a consequence, Indian think tanks had turned their attention to foreign funders.

For leading global Indian intellectuals such as Jagdish Bhagwati, writing for the Times of India, this should be a source of serious concern for India (Bhagwati, 2010):

> *Funding does constrain what you will do: this is simply a matter of prudence, not of being "bought". I will give one personal example. I was on the board of an important Indian NGO which deals with trade issues. This NGO was fully sympathetic to myriad writings by me and professors Arvind Panagariya and T N Srinivasan, among others, warning how the demands to include labour standards in trade treaties and institutions were tantamount to "export protectionism" (in the sense of seeking to raise the cost of production abroad to moderate competition). We had forcefully argued that these demands must be rejected as being driven by labour unions in the West, which were wrongly fearful of trade with the developing countries.*
>
> *Having been funded by foreign agencies which wanted them to work with foreign think tanks, the Indian NGO had organised a seminar on the subject in Washington DC, under joint auspices with Carnegie. It wanted me to play a prominent role, but it had to agree to my being downgraded because Carnegie had embraced the protectionist agenda on labour standards. With foreign funding, both current and prospective, the Indian NGO felt that it had little choice and sought my indulgence. I resigned over the incident from the NGO, only to return later as I saw the difficulty in which foreign funding had placed its able director. He had integrity; he was penitent. But he had to be prudent or his NGO would be financially crippled.*

He argued for a more concerted effort by the Indian government and Indian foundations and philanthropists to support local think tanks and experts. What Dr. Bhagwati was describing happens elsewhere, as well. I once found an account by Teresa Hayter on the fate of a report she produced a long time ago for the Overseas Development Institute, and that was critical of the World Bank, particularly interesting as it described precisely what Dr. Bhagwati feared.

Anuraag Sanghi contributed to the debate with an article named *Collusion or collaboration? The Think Tank Initiative* (Sanghi, 2010). Mr. Sanghi was critical of the role that foreign funders play in the development of India's think tank community, particularly because of the intellectual and ideological constraints their involvement implies — echoing Dr. Bhagwati's concerns.

In response, Suman Bery, director-general of the National Council of Applied Economic Research, and a member of the Prime Minister's Economic Advisory Council, published an article on the need to diversify think tanks' support. Suman Bery also provided insights into the challenges that the NCAER's business model faces (Bery, 2010):

> *[The NCAER was originally] expected to support itself through contract research for at least two reasons: first, because there was no other funding model available and, second, to ensure that its work programme addressed practical problems rather than reflecting the intellectual interests of its staff.*

However, this is no longer possible as the bureaucratic control that the state effects on the council has reduced the space for intellectual and professional development. He called, as a consequence, for local corporations and foundations to take on the responsibility of funding research.

All these commentaries and a recent study by Srivastava (2012) award a significant degree of importance to the state–society settlement achieved in post-colonial societies in the region. Given the region's relatively recent independence from Britain, this presents an important point of difference with the Latin American and Southeast and East Asian experiences. There, colonial and post-colonial periods certainly had an effect, as they would be expected to, but, at least in the case of Latin America, these have been felt over a much longer period of time and so have had to share the limelight with other more recent and overlapping factors. In the Southeast and East Asian case, the post-war period appears to mark a significant turning point for the region as a whole.

For Srivastava, the key to understanding think tanks in South Asia is to understand the nature of the State and how it has become embedded in society and all its expressions, including think tanks. To do so, he employs Michael Mann's distinction between despotic and infrastructural power in India, Pakistan, Bangladesh, and Sri Lanka. Despotic power introduces an interesting component in the relationship between think tanks and the State. It refers to the actions the State can undertake without consultation or negotiation with its citizens. In the specific case of think tanks, all South Asian States (this is in fact the case in all other regions) have developed and enforced a wide range of regulations to control the activities of civil society organisations. These regulations can be directly targeted towards them, for instance, establishing conditions on registration, allowing certain types of activities, special tax structures for non-for-profit organisations, etc. They can also be indirect by limiting tax incentives to provide donations to non-for-profits, labour legislation that affects the way in which think tanks can hire new staff, procurement policies, etc.

Infrastructural power refers to a more reciprocal relationship between society and the State, which is best illustrated by the different types of regimes found in the region — democracy, dictatorship, authoritarian, or semi-authoritarian — and the degree of pluralism and space for deliberation in society that is possible in each.

Inevitably, then, think tanks are not simply bridges between knowledge and politics (or between any other two or more groups or communities), but part of the same state–society field, to use Medvetz terms, playing different roles depending on the space available to them. By and large, Srivastava, as Bhagwati and others, concludes that think tanks follow the demands of those in power and who are capable of expressing their agency (whether they be the State or donor agencies).

The collective South Asian experience offers an abridged history of think tanks "after independence," which presents a parallel to that of Africa. Srivastava identifies two main post independence moments. The first is during the early period of independence when governments, in an expression of nationalist aspirations, developed a stronghold on policy-making and, as a consequence, the civil service became the gatekeeper and the only channel of any information intended for decision-makers. This key role is still present today and has the effect wherein think tanks are frequently located in the capitals, close to those in power, and largely dependent on their governments' own agency for funding and defining their research agendas.

Such a strong State and, probably even more importantly, an empowered civil service have influenced the development of think tanks by defining the terms of their development and keeping them close to the State (both physically and intellectually). According to Srivastava, in the case of India and Sri Lanka, the relative openness of the political and economic system has encouraged the development of more think tanks with different research agendas and interests than found in Pakistan and Bangladesh. But another factor may explain this difference. At least in the case of Bangladesh, the State enjoys relatively little autonomy from international discourses, and the international donor community plays a highly influential role in setting the research agenda and directly funding research and think tanks.

The second period is characterised by a retreat by the State from funding social sciences, explained in part by a pursuit of technology-driven growth, and the emergence of donor funding as an alternative, first led by the Ford Foundation and then, as in Latin America, by a host of others: for example, the Rockefeller Foundation, the Canadian International Development Research Centre, the Konrad Adenauer Stiftun, USAID, and the World Bank. The Think Tank Initiative led a more recent wave of foreign investment.

As the State retreated and international donors increased their presence and importance, think tanks' business models changed, as well: from academic models to contract models. The type of funding provided by donors, in line with current trends, has tended to limit rather than open the space for think tanks to develop and operate, placing little emphasis on long-term research and instead focusing on what is termed policy-relevant research: "short-term, immediate, instrumental, micro-level and incremental knowledge rather than critical and theoretical research. Such a historical approach and inductive research leaves both the past and the future from the ambit of analysis, leading to a constrained research agenda" (Srivastava, 2012, p. 115).

Think tanks' attention has also shifted from local politics to international politics, as it is the latter that more directly influenced their future funding prospects and research agendas.

The Indian State, relatively more democratic than its neighbours, has made more use of infrastructural power in relation to think tanks, thus creating (or allowing) greater spaces for their development supported by foreign and domestic private funders. The absence of military rule and the fairly decentralised nature of the Indian State have allowed more think tanks, with different interests and publics, to emerge. On the other hand, the advent of military or autocratic rule in Bangladesh and Pakistan brought forth a more despotic use of power and a clear control and limiting of think tanks' space. In the case of Sri Lanka, the limiting factor was not authoritarianism but the development of a large welfare state that effectively crowded out civil society and limited the space for think tanks to operate in.

Srivastava hence concludes that South Asian think tanks are not indigenous to the region nor have they developed organically: Anglo-American templates, experience, and support have driven much of the inspiration and agency in the sector. This places a greater emphasis on the influence of external forces that either the Latin American or Southeast and East Asian experiences account for.

In Latin America, it is unlikely that the recent boom in the think tank community would have taken place without significant investments from international donors. However, given the longer tradition of policy research, critical thinking, and a longer post-colonial period, it is possible that Latin American think tanks have by and large been able to resist international development discourses more than their peers in South Asia and therefore maintain a more domestically (or at least regionally) driven intellectual agenda. Regional centres such as ECLAC and FLACSO, and the presence of regional funding bodies such as the Inter-American Development Bank and the Andean Development Corporation (both of which fund think tanks through a mix of core, programme, and contract mechanisms — even if they do not have a think tank funding policy), have provided a balance to truly external influences.

Similarly, although dictatorships in Latin America by and large limited the space for think tanks, the critical mass of researchers and intellectuals that existed found, by necessity, new spaces in which to operate during difficult years of despotic power. In turn, this allowed for smooth transitions into periods of infrastructural power.

Nonetheless, I would not go so far as to suggest that there is nothing organic in the South Asian experience. The differences in the use of despotic versus infrastructural power found in each country do account for the influence that domestic forces have had in the development of think tanks. The emergence of think tanks close to the State at the outset of the independent period is also evidence of the organic nature of their emergence. And the role of foreign donors in shaping modern think tanks is greatly due to the State's own withdrawal from the scene.

Additionally, the story is far from over. A recent turn towards authoritarianism and the increasing closing of the civic space have seen think tanks in opposition to the Modi government struggle to access funding and secure spaces for influence. India's relatively decentralised political system and the emergence of a new class of local, politically minded philanthropy help explain the resilience of the Indian think tank community.

Sub-Saharan Africa: Foreign Influence

Given the attention from donors, it is surprising that the region does not yet enjoy the same degree of research on think tanks that others have. The focus is more on good governance and civil society organisations, particularly those who claim to speak on behalf of the poor (Broadbent, 2012). Think tanks continue to be seen as a different kind of civil society: elitist, staffed with economists who work directly with or for governments, and whose control of policy processes can sometimes be interpreted as being in opposition to the interests of the poor. To unpack this, Kimenyi and Datta (2011) focused their analysis of sub-Saharan African[2] think tanks on two political dimensions: the politics of State power and the politics of external influence. They used these to explore the development of think tanks across four distinct historical periods — colonialism, post-independence and single-party rule, authoritarian/military rule, and political and economic liberalisation — which, although did not happen at the same time in every country, can be found in most.

The politics of state power describe the extent to which policymaking is concentrated or dispersed within society. It permits us to distinguish between periods of single-party, military, and semi-democratic rule. Unlike Latin America, where power in politics appears to have been historically organised across class lines (with some exceptions, for example, the case of post-Evo Morales's Bolivia, where identity politics have generated an interesting shift in the role of political parties and think tanks, Loayza, 2011), Kimenyi argues that in Africa, divisions are ethnic or religious — even if more recently these are blended with generational and class differences. As a consequence, leaders have employed these differences to rally support for themselves and their policies.

The politics of external influence are expected to be of significant importance in Africa. International political, economic, and military influences, and more recently development cooperation, have been all pervasive in Africa. External influences reflect the control that foreign players and interests have had over policymaking: from colonial authorities, development experts, development agencies, non-governmental organisations, etc. Kimenyi's and Datta's analysis of African think tanks' historical development provides an interesting opportunity to deploy the four

---

[2]They do not consider South African think tanks which would, given their unique context, require a separate study; not attempted here.

different perspectives or schools of thought briefly mentioned earlier. From a clear control of the colonial State, African politics came to be led by elites in the short period after independence, then the State again during long periods of authoritarian and military rule, followed by the pluralism of political and economic liberalisation since the 1990s, and finally a new focus on the organisations themselves as the international development agenda pays increasing attention to evidence-based policy and the efficiency of aid flows.

Through the lens of this framework, Kimenyi and Datta have described the shifting roles that think tanks (which they consider to include independent think tanks, research centres affiliated to the state or universities, academics, foreign experts and institutions, political parties, and civil society organisations more generally) have played. They found the origin of modern think tanks in Africa well within the colonial period when their administrations, far from concerned with the development and promotion of policies to help the African population, set up centres such as an experimental botanical research station in Lagos (1893), the West African Institute for Social and Economic Research (1950), and several East African institutes focused on agriculture and livestock, to support colonial rule. Clearly, their emphasis was on the study of methods to improve the production and export of strategic crops such as palm oil and rubber, as well as mechanisms for better ruling the colonies. Much as in the case of China and Vietnam, the role of think tanks was to support and legitimise the ruling regime. However, there were also instrumental think tanks similar to those set up in the United States in the early 1900s, with the purpose of supporting the implementation of policy.

During the initial post-colonial period, these colonial research centres were reconfigured to serve domestic interests, as in the case of South Asia, where independence and state control were seen as the embodiment of nationalism. Significant investments in planning ministries were accompanied by the formation of new research units both within the State and in universities. Many of these organisations were led or influenced by the intellectuals who had been responsible for agitating for independence and influencing the political ideologies of the newly independent nations.

In Britain, this re-focusing of African think tanks was met with a renewed interest among researchers to influence the development of policies in and towards the newly independent nations. The Overseas Development Institute (ODI), for example, was founded in 1960 with funding from the Ford Foundation "to be a centre for work on development problems, carry out studies on development topics, be a forum for those concerned with development, spread information, and keep the urgency of the problems in the eye of the public and decision makers."[3]

The new elite's ideology, a rejection of capitalism as a form of colonialism, came to influence African politics during the first years of independence in the 1960s and 1970s. During this time, the relationship between academics and policymakers was cordial and constructive.

Soon, however, single-party rule turned into authoritarian and military regimes that developed alongside a widening rift between politicians and academics. Post-independence leaders had focused on consolidating power and were not interested in the emergence of a strong intellectual class, nor were there significant overlaps between the *freedom fighters* that had won the independence struggle and the intellectual elite. Unlike Latin America, where many of the *libertadores* and those directly involved in the independence struggle in the late 1700s and early 1800s were part of the same intellectual elites that took on the leadership of the new nations, often educated in Europe, in Africa, the *freedom fighters* were largely uneducated and did not appreciate intellectuals' role and criticism (Mkandawire, 2000).

Excluded from the state, intellectuals turned to civil society and foreign funding, thus becoming increasingly distant, in both a literal and figurative way, from policymaking and dependent on foreign interests and funds. This reliance on foreign funding was accentuated as the military regimes and authoritarian governments cracked down on all intellectual and ideological dissent, particularly in universities. It is within this context, not unlike in Chile, that modern African think tanks would later emerge.

Throughout this period, however, African leaders relied on foreign ideologues and technocrats — both for affirmation and expertise (Kimenyi and Datta, 2011). The economic crisis of the 1970s and 1980s led to an imposed economic and political liberalisation in many African countries. The structural adjustments, in turn, led to a sudden opening of the policy space. However, unlike other regions where local organisations and experts were able to meet the demand for expertise and new ideas (with some foreign players too), in Africa, liberalisation and the increased donor support it brought were conditional upon the use of foreign expertise.

---

[3] From ODI's website: http://www.odi.org.uk/about/50years/#timeline.

Kimenyi and Datta report that in the late 1980s, there were 100,000 donor-funded expatriate advisors working in the public sector of 40 Sub-Saharan African countries. More generally, the liberalisation process implied important effects for the development of think tanks in Africa.

Initially, the number of policy research centres flourished. Their agenda was heavily influenced by donors' own agendas — first political and economic liberalisation, trade and economic integration, and then poverty reduction, human development, aid architecture, and budget tracking. The monitoring or auditing function of think tanks was strongly emphasised as their funders saw think tanks as a mechanism to protect their own investments rather than an institution valuable in its own right. It is not surprising that the World Bank has been instrumental in the establishment of many think tanks in the developing world, and anywhere where it has had an important role in funding and promoting economic reform. In Africa, ACBF was for decades a preferred mechanism to support the establishment and development of economic policy think tanks like ZIPAR throughout the region.

This final point has further implications. It often placed think tanks (and more specifically academic research centres and non-governmental organisations) at odds and in direct conflict with their governments, who are often seen as the sole barriers for progress, and drew them closer to an international or regional policy and research community than would be expected of think tanks in any other circumstances. This, in turn, reduces the possibilities for formal interaction with political parties or platforms. At the same time, as the discourse moved away from the Washington Consensus, think tanks with strong ties (both historical and current) to the Bretton Woods institutions and their ideas fell foul of international non-governmental organisations and donors promoting alternative, post-Washington Consensus ideas. At a meeting of civil society organisations to discuss tax justice in Zambia, I asked a small group of participants about their views on think tanks in Africa. I was surprised by their response. The general view was that think tanks undertook academic research, mostly, if not solely, on economic policy issues, and were close to, if not part of, the government. But most importantly, they were seen as fundamentally pro-economic liberalisation.

In hindsight, this should not have been surprising at all. The most important efforts to support the development of think tanks in the region have coincided with this assessment. ACBF is almost entirely focused on promoting economic policy research, and it has a clear academic emphasis. The African Economic Research Consortium (AERC) also supports economic policy research. The Canadian IDRC, which has been a strong supporter of think tanks, has also focused most of its efforts on economic and social policy think tanks. The Secretariat for Institutional Support for Economic Research in Africa (SISERA), as its name suggests, supported economic policy think tanks, many of which were then part of the Think Tank Initiative's African cohort.

More recently, think tanks like the African Center for Economic Transformation (ACET) in Ghana, the South African Institute of International Affairs (SAIIA) in South Africa, the African Center for Equitable Development (ACED) in Benin, the Kenya Institute for Public Policy Research and Analysis (KIPPRA) in Kenya, Nkafu Policy Institute in Cameroon, and, even, the Policy Centre for the New South (PCNS) in Morocco, are driving the formation of new cadres of think tankers and think tank communities convened around projects, conferences or training opportunities.

The result is a new wave of think tank formation, development and innovation across the region.

The African experiences then suggest a model in which the State and foreign forces have competed for power and control over intellectual agency. Although independence initially opened the political space, this was swiftly closed by authoritarian rule. Liberalisation, on the other hand, might have brought forward new opportunities for think tanks to develop, but foreign interests greatly limited their room for manoeuvre right from the start, conditioning not only their own development and functions but also their relations with other domestic players.

Bringing Back Politics

In this section, I have outlined the different frameworks employed and intended to make politics explicit in the study of think tanks in developing countries. They denote a shift in focus away from the organisation and to its environment. Table 2 provides a summary of the dimensions of analysis of the studies discussed earlier.

Studying the rise of think tanks from an explicitly political lens offers opportunities that an organisational perspective alone cannot. Josef Braml's comparative study of think tanks in the United States and Germany provides a good example of the nuances that can be drawn from studies such as these (Braml, 2004). Work by Braun *et al.* (2007), who

Table 2.   Dimensions of analysis per region.

Region	Dimensions
Latin America	Degree of institutionalisation of the political system and political parties
	Degree of institutionalisation of the relationships between political parties and external think tanks
Southeast and East Asia	The politics of production
	The politics of power
South Asia	The exercise of despotic power by the state
	The exercise of infrastructural power by the state
Africa	The politics of state power
	The politics of external power

wrote about exogenous and endogenous factors for think tanks across the developing world, and work by Abelson (2006), Smith (1991), and Ricci (1993) (although the latter is more historical in nature) are greatly illustrative in this respect, as well.

Braml's framework, just as the studies for Latin America, East and Southeast Asia, South Asia, and Africa described earlier, employs a relational (organisation *homo mediaticus*) and functional (knowledge producers, transmitters and interpreters, and conveners, networkers, and influencers) description of think tanks, both dimensions forcing him to look outside the organisation. His focus on relations is emphasised by choosing *affiliation* as the main differentiating characteristic for his classification of think tanks: ideologically identifiable advocacy and party think tanks, and ideologically non-identifiable academic and contract think tanks.

Braml considered a number of contextual factors or forces that affect the formation, development, and functioning of think tanks to be relevant to the literature on think tanks globally. Within each, he paid particular attention to the opportunities and barriers that they offer. In essence, he assumes that the State plays a key role in both Germany and the United States — far more than pluralists would in the case of the latter.

For example, he unpacked the institutional environment in which think tanks have developed in the United States and Germany by focusing on the specific characteristics of their political systems and how they affect the mandates and roles of think tanks. He considered how differences between Presidential and Parliamentary systems — and the nature of the divisions and balance of power — increase or reduce opportunities for think tanks under certain conditions. Political parties both in government and in opposition, as in the Latin American cases, are given particular attention: He found that stronger and more cohesive parties, as the ones found in Europe's parliamentary systems, provide fewer opportunities for policy entrepreneurs than the more loosely managed and defined political parties in the United States.

Similarly, he argued that centralised States, such as those in South Asia and East and Southeast Asia, reduce the space available for think tanks while decentralised ones create opportunities at various levels. This is consistent with the Argentinean and Indian experiences, for instance, where the nature of the state has encouraged and supported the formation of think tanks at the provincial and state levels, respectively. It also explains how foreign funding that creates additional markets or spaces for think tanks to appear even in contexts where the State and policymakers are not supportive of them.

The nature of the policymaking bureaucracy also offers insights into the opportunities for think tanks. Like Srivastava, Braml awards civil servants a critical role in defining the roles played by think tanks.

In general, this line of analysis seeks to identify the political opportunities that exist for think tank formation and development. Abelson (2009), Weiss (1977), Smith (1991), Weaver (1989), Rich (2006), Naughton (2002), Tanner (2002), Belletini (2007), Stein and Tommasi (2008), Li (2009), Garce (2018), and others have all, directly and indirectly, conferred the State and the political process a great deal of responsibility in shaping the formation and development of think tanks.

In their analyses, a number of factors that reflect the importance of institutions in the study of think tanks are continuously identified as opportunities or barriers:

- The degree of *porosity* (i.e., the opportunities for third parties to participate in the workings of the party) of political parties and the party discipline/independence of elected officials, and similarly that of the civil service combined with its professionalism;
- The degree of complexity of the national and subnational government — number of government departments, their competencies, policy capacities, etc.;
- The degree of separation of power between the executive, the legislative, and the judiciary — and the level of competition and balance of power that exists between and within them;

- The pervasiveness of the state — though legislation or influence — in the private sector and civil society;
- The role of certain actors of the policymaking process in the path of a policy and the length of the policy process, which in developing countries must affect the roles played by foreign donors, civil society, and consultants;
- The length and nature of electoral processes;
- The role of the media in demanding information from other political players;
- The acceptance or degree of value that is conferred to evidence in the policymaking process; and
- The influence that foreign actors have on national and subnational politics and policy processes across all the factors described earlier.

The potential for analysis is rich and helps to further explain think tanks' own internal functioning. For example, the arguably "porous" nature of the civil service in the United States,[4] as opposed to the British or German ones, for example, means that the movement of experts from think tanks into policymaking positions — and vice versa — is quite common, as is, inevitably, its politicisation. This positions think tanks as an attractive vehicle or stepping stone for a career in policymaking. The weaker and less cohesive nature of political parties in the United States also explains the higher demand for think tank ideas in its Congress, where individual members are freer than their peers across the Atlantic to put forward their own legislative initiatives and vote independently and against their party. Finally, the clear separation of powers between the legislative, executive, and judiciary and the dynamics of Federal- and State-level politics create ample opportunities for think tanks to engage with the State (Braml, 2004).

Certain elements that resonate with this analysis of think tanks in developed countries can be drawn from combining the cases described in the previous chapter and this chapter's more explicitly political and systematic analysis:

(1) It is interesting to note that all the studies and their findings have awarded a prominent role to the State (and its relationship with other

---

[4]Up to 10,000 new people can be appointed by a new President (Abelson, 2006).

political players) in defining the space in which think tanks can originate and develop. Far from being a barrier to their formation, strong and even authoritarian States are responsible for promoting them both directly and indirectly. It appears then that democracy helps (and it can explain their number and ideological diversity), but it is not a necessary factor for their existence. A State that values research may very well prove to be an excellent and more efficient and effective promoter of think tanks.

A possible reason why democracy does not appear to play as important a role as expected is that political and economic reforms have not necessarily come hand in hand. In Latin America and Africa, the harsh economic reforms of the 1980s and 1990s were often implemented before strong democratic institutions had developed. In Southeast and East Asia, evidence of the presence of think tanks in starkly different democratic and non-democratic contexts as Japan and China, for example, does not necessarily correspond to a clear-cut linear relation between openness and think tanks.

Economic reform, however, demands support and advice, and think tanks have offered an excellent vehicle for both. The economic liberalisation promoted by the World Bank and other regional development banks was accompanied, and legitimised, by funding for research and think tanks, more specifically.

(2) Much more important, in fact, is the degree of institutionalisation of other political players that make up the political, economic, and social elites of a nation. The State, the Church, the media, the corporate sector and philanthropic bodies, donors, academia, and others play important roles vis-à-vis think tanks' formation and development, even in the absence of democracy.

(3) Funders and "development agencies," particularly foreign, have hence played important and different roles in each region — to a greater and lesser extent depending on their overall influence in national politics — and may help explain this rather surprising idea that think tanks can thrive even in non-democratic or politically closed societies. The case of India described earlier is paradigmatic and reflects a similar pattern observed in Latin America and Africa: When the State retreated and withdrew funding from the social sciences, foreign funders moved right in. As a consequence, even during the heyday of military rule and economic crisis, research continued to be funded.

Dependence on the State and foreign agencies for funding for research, however, appears to explain a relatively homogenous think tank community in certain disciplines (e.g., economics) and locations (e.g., Sub-Saharan Africa.) The presence of non-State domestic funders, on the other hand, is more likely to lead to greater ideological and political plurality among think tanks (e.g., Chile, India, or Indonesia).

Therefore, foreign agency funding has an effect on the relations between think tanks and other domestic political actors. In Latin America, weak political parties, States, and private foundations have sometimes been replaced by foreign funders. In Africa, foreign influence is of such importance that it rivals the power of the State itself. In South Asia, the influence of foreign funders has generated a public debate in India, where there is concern of the potential of losing intellectual autonomy. In East and Southeast Asia, concern for foreign threats, rather than funding, has been a source of motivation for the State's investment in think tanks.

Of course, the effect this has on think tanks' legitimacy differs too. In Peru, foreign funding is not necessarily an issue in itself — it all depends on the eye of the beholder. For mainstream economists, funding from NGOs such as Oxfam, which are closely associated with social movements opposing the privatisation of services and the expansion of mining exploration, is seen with suspicion. For many human rights NGOs, on the other hand, the funding these economists receive from the World Bank or the Inter-American Development Bank is a source of concern.

The lively, albeit unexpected, debate that the Think Tank Initiative's launch generated in India is a testament to the concern over foreign influence in the national think tank community. In the years that followed, it may have created a greater interest in think tanks from political leaders and local philanthropists.

The issue matters. As Broadbent found in the study of the debate on genetically modified organisms in Zambia, JCTR's position was often put into question not due to its religious undertones, or its evidence, but because of its foreign funding (2012). This perception of undue influence from foreign funders has been used increasingly across the world to challenge the legitimacy of think tanks.

(4) There also seems to be a both a historical progression and cycle at play, which adds weight to the idea that think tanks develop in waves:

from elite and state control, to more pluralistic contexts, and back. Think tanks are not experienced by all in the same way. Periods of great ideological upheaval (pre- and post-independence, for example) were followed by periods of technocratic rule and the concentration of power with new elites or the State. This, in turn, has been followed by economic and political liberalisations that, with varying degrees of foreign influence, have introduced new ideas and resources to the think tank community and affected their relations with domestic institutions. Inevitably, where non-state domestic funding has not been present and political competition has been absent, this has in fact reduced the plurality expected of a more liberal political space, as in the case of Chile's (Cociña and Toro, 2009) and Colombia's (Londoño, 2009) coalition governments or in Zambia. Therefore, the development of think tanks needs to be seen in a historical perspective, and this in turn should not be perceived as linear.

This has been a long introduction to discussing the elitist, statist, and pluralist perspectives as they play out in the study of think tanks in developing countries. In the following chapters, I will explore these various perspectives in greater detail and delve deeper into the effect that think tanks' own external environments have on them.

Chapter 7

The State as the Main Sponsor of Think Tanks

Applying a political lens to the study of think tanks awarded the State a privileged role in their development, both in promoting and stalling it. Naturally, the statist perspective is particularly attractive for scholars of strong governments and rather authoritarian countries. This view is sometimes quickly dismissed in the literature from developed countries, for it does not agree well with the idea of pluralist and democratic societies, nor with the good governance agenda — although comparisons between the United States and Canada (Abelson, 2009), Britain, and Germany (Braml, 2004) emphasise the relatively more statist nature of policymaking in the latter cases. It is often assumed, and there is certainly some evidence of this in Latin America, Africa, and South Asia, that political liberalisation (or the strengthening of democratic rule) leads to an increase in the number of think tanks. However, this perspective provides an alternative account, suggesting that, in fact, liberal democracies, while certainly helpful, are neither necessary nor sufficient for the emergence of think tanks, and that these may thrive even within authoritarian and military regimes.

The history of Chinese think tanks, described in this chapter, offers an opportunity to understand this dynamic and unpack the roles that they can play even in apparently closed political contexts. In general, the emergence of think tanks in Southeast and East Asia more broadly is linked to the role of a strong developmental state player: for example, the State in China and Vietnam, the Private Sector in Korea and Japan, and Civil Society (represented by prominent leaders, similar to the *caudillos* of

Latin American) in Indonesia and Malaysia (Nachiappan *et al.*, 2010). In all cases, they have founded and patronised the development of think tanks — which, as a consequence of their patrons' differences, are themselves quite diverse.

Although statist views would consider think tanks' influence to be marginal and limited to legitimising the State, the Chinese case provides evidence that think tanks can thrive in non-democratic contexts and that, ironically, they can be a force for political and economic openness (Zufeng, 2009; Shambaugh, 2002; Naughton, 2002; Gill and Mulvenon, 2002; Glaser and Saunders, 2002) rather than mere tools of the State. In China, unlike in Eastern Europe or Chile, these think tanks have been funded by the very system they have helped to steer.

In recent years, a new wave of think tanks in the North Africa and Middle East region has confirmed this. Vision 2023 in Saudi Arabia mirrors President Xi Jinping's call for more "think tanks with Chinese characteristics," which saw a boom in think tank formation in China in the mid to late 2010s. The Policy Centre for the New South in Morocco has emerged to play a role in reshaping relations between state, social, and economic actors in the greater Atlantic region — greatly supported by the Moroccan government.

The Chairman Also Needs Advice

There are various examples of think tanks emerging during periods of authoritarian rule. In Latin America, they did so and operated during military dictatorships — Chile being a case of particular interest as this saw the rise of a political opposition as well (Puryear, 1994). Kimenyi and Datta, for example, describe how think tanks first emerged in Africa during the colonial period to support the policymaking of the colonial administrators (Kimenyi and Datta, 2011). In South Asia and in Southeast Asia, economic and social policy research centres have continued to develop in semi-democratic and autocratic contexts (Srivastava, 2012). The nationalistic governments that took power after independence set many of them up. In these cases, as the ruling regimes withdrew from supporting the formation of new centres or the operations of existing ones, foreign donors stepped in. Not in China. There, the State has played a direct and continued role in the development of think tanks — albeit not always a positive one.

Chinese think tanks are not new, and the country's most recent economic and political reforms alone cannot explain why their numbers have

expanded in the last three decades (Shambaugh, 2002). Most first-genera-
tion Chinese think tanks have their roots in the 1950s and 1960s
(Shambaugh, 2002). Interestingly, the first attempt to establish a think tank
outside the State apparatus was as early as 1956. According to Tanner
(2002), concerned that his advisors had failed to predict the political unrest
in Hungary, Poland, and the Soviet Union, Chairman Mao instructed
Premier Zhou Enlai to establish the Institute of International Relations
(then called the Institute of International Studies) under the Ministry of
Foreign Affairs. That same year, the Bureau of World Economics was set
up, and the Chinese Academic of Sciences and Philosophy and the Social
Science Department established an economics research institute with an
interest in international economics. By 1958, it further adopted a focus on
world politics.

Driven by a strong nationalism strand (Nachiappan *et al.*, 2010), com-
petition with Moscow for global influence also played an important role
in China setting up of a number of other centres including: the Afro-Asia
Research Institute (1961), the Soviet-East Europe Institute, the Latin
America Institute, the India Research Institute, and the International Law
Institute (1963). In 1964, the College of Foreign Affairs and the First
Foreign Languages Institute were set up. Finally, a number of other cen-
tres were established within Chinese universities.

This first generation of think tanks was heavily influenced by the
Soviet research model: Their analyses had to support policy, and not
challenge it (Shambaugh, 2002). It is clear that a concern for protecting
and legitimating the system (both domestically and internationally)
played an important role in the decision to set them up (Nachiappan *et
al.*, 2010), thus confirming the legitimising functions we all expect to
find in China. However, this motivation is also not too different from
that of the German State, which recognises the importance that these
centres have for their policymaking responsibilities and provides most
of their funding (Braml, 2004), or for the various think tanks historically promoted and funded by the Federal Government in the United
States on issues as diverse as military, foreign, and economic and social
policy.

The Cultural Revolution, however, led to the closure of most of these
centres and the emergence of a new system. In 1977, the Chinese
Academy of Social Sciences (CASS), a *think tank of think tanks* that
served as the model for the Vietnam Academy of Social Sciences (VASS),
was founded (Tanner, 2002). The research model in the 1980s had
changed, and the advisory and legitimising functions of think tanks were

replaced by a more reformist agenda. Several of the second-generation centres that emerged were set up under the auspices of key party leaders, particularly those with a reformist orientation, who believed that the advice they were receiving from traditional Party Departments and State Ministries was inadequate for the rapid economic reform that was needed (Tanner, 2002) as well as for their own political interests. China's rapid entry into the global economy demanded more empirically based policy options — rather than ideologically based — and so the think tank community that developed had a much greater appreciation for the interaction between domestic and international policies, the political economy of change, and globalisation, as well as a much deeper understanding of international organisations and their roles in world politics and processes (Shambaugh, 2002).

Interestingly, distrust by politicians of the system meant that these new think tanks were established under direct patronage of individuals rather than the organisations or institutions they represented — all the way to the top: For instance, Premier Zhao Ziyang purposely patronised independent think tanks to access ideas from outside the party (Naughton, 2002). They were quite personalised and often no more than the consolidation of informal group consultations among a few key policy intellectuals and senior leaders. This ambiguous relationship between leaders and intellectuals was not a coincidence and, as Tanner argued, it allowed the centres to be rather innovative and independent — they were safe from political reprisals because the centres were as easy to set up as they were to shut down.

Parallelling the history of think tanks in the United States, these centres represented the *Salomon Houses* of reformist leaders who, within the system, sought to promote certain ideas. This is remarkably similar to the manner in which Colombian political leaders used think tanks and political publications to gain intellectual clout over their peers (Londoño, 2009), or the role played by a more recently founded think tank, the Centre for Social Justice in the United Kingdom, which was set up by former Conservative Party leader Iain Duncan Smith as a way to promote new ideas within the party (Snowdon, 2010). In fact, Shambaugh's view is that while many of the original Chinese think tanks were like the stepchildren of the Soviet system, in the 1980s, they were redeveloped under Anglo-American models (Shambaugh, 2002): adopting more proactive problem identification and influencing functions.

The third generation of think tanks emerged after the Tiananmen Square protests and crackdown of 1989 (Tanner, 2002). The State repression that followed the protests initially devastated many of the economic and political reformist think tanks that had emerged during the 1980s. Just as in the aftermath of the Great War, the Great Depression, and the Second World War in the United States, the military coup in Chile, or the fall of Suharto in Indonesia, eventually, the new think tanks' attention turned towards understanding the social, economic, and political causes of the unrest. This reflected a renewed concern for threats to the system (Nachiappan *et al.*, 2010) and the recognition that think tanks may be invaluable for it and for policymaking, after all.

The main characteristic of the third generation is the *marketisation* of researchers' employment and a reduction in think tanks' dependence on ministries and political patrons (Tanner, 2002). According to Cheng Li, the convening of a tripartite of officials, entrepreneurs, and scholars marked the emergence of the new think tanks. These, like the China Centre for International Economic Exchanges, the Chinese Economists 50 Forum, and the China Centre for Economic Research at Peking University, have been partly strengthened and promoted by the large numbers of "returnees" who, after decades of scholarship and work in the West, are findings increasing opportunities in a more open Chinese State and its growing think tank community (Li, 2009):

> The most notable [of several important developments] is that three distinct groups of elites — current or retired government officials, business leaders, and public intellectuals — have become increasingly active in promoting their personal influence, institutional interests, and policy initiatives through these semi-governmental organizations. In present-day China, think tanks have become not only an important venue for retired government officials to pursue a new phase in their careers, but also a crucial institutional meeting ground where officials, entrepreneurs, and scholars can interact.

The relationship between the State and think tanks in China shows a remarkable change in a relatively short period of time (Li, 2009). Whereas, traditionally, think tanks would have been given space (or had to acquire it) as a result of the patronage of powerful officials, or the interest of Premiers and Party Chairmen, today, think tanks have become a place to reflect on a long public career or plan a future one — clearly adopting the Anglo-America revolving door model. Their relation to power is therefore more plural, open, and even critical (within acceptable boundaries). But what is by far most interesting is what appears to be a concerted, and historical, effort by the Chinese State to promote the formation of think tanks. At an international symposium of think tanks in China in June 2013, the participants, mostly Chinese think tank leaders, policymakers, entrepreneurs, and scholars, reached a clear consensus on the purpose of such a gathering. On the agenda was a discussion, not on whether China needs more think tanks, but on how the think tank community can be strengthened and expanded. Think tanks are seen as key players in Chinese development plans.

In 2013, President Xi Jinping "for a new type of think tank catered for China to 'promote scientific and democratic decision making [...] And strengthening China's soft power', the country has seen an exponential increase in think tanks" (Mendizabal, 2016). This led to the sudden formation of think tanks in a brief period of time. Much of this growth was explained by the relabelling of existing organisations — a perfect example of the political nature of self-labelling as described by Medvetz.

But a critical aspect of this most recent wave of think tank formation is that think tanks were not simply called to help inform Chinese policymaking but, crucially, global narratives about China. If before, think tanks were described as "windows" into other policy communities, now the emphasis was on super highways to promote Chinese interests abroad (Mendizabal, 2012). Not surprisingly, this State- and Party-sponsored campaign for new think tanks coincided with the launch of the Belt and Road Initiative in 2013.

Economic liberalisation and China's economic and political development have also made a difference. Today, think tanks have more clients and prospective think tank researchers have more career choices than before. As a consequence, researchers move to and from the public and corporate sector, using think tanks in much the same way as they are used in the United States, Britain, or Germany: as stepping-stones (Naughton, 2002). Again, however, the State has played a key yet indirect role. Firstly, by redefining the prevailing worldview that is increasingly consistent with Western economic paradigms and devoting significant resources towards it, it has attracted new policy entrepreneurs from abroad. Secondly, policymakers have turned towards competition between think tanks by tasking them to work on the same issues and hence provide opportunities for relatively independent thought to emerge. Patronising a single think tank is no longer cost-effective. And thirdly, the development of regional policy communities in Shanghai and Guangdong, for example, has created new opportunities for think tanks at the sub-national level (Naughton, 2002). This is particularly interesting as it suggests that think tanks are called to play a role in supporting their cities' efforts to carve a space for themselves at the national level. The decentralisation of power (even if this is still kept within the control of the single party) thus offers think tanks new opportunities to develop.

Hence, although there is an undeniable role for the economic and political liberalisation that China is undergoing, the agency of the State, whether it is by directly setting up new centres or creating the space for private initiatives, remains central to the development of the think tank community. Even unintentionally, the State has played a key role, as in the case of the Chinese policy to forcefully retire officials, creating a growing number of restless seasoned and tenured candidates for think tank leaders and fellows (Li, 2009).

Public Research Funding

As might be expected, the State's control over think tanks is closely linked to its funding of them. Kimenyi's and Datta's study of think tanks in Sub-Saharan Africa stressed the tension between the power of the State and donors, and this is closely related to who provides their funding (2011). Srivastava's study of think tanks in South Asia also showed that there is a trade-off between State and foreign funding: When the State withdrew from funding social and economic research, foreign funders moved in (2012). Implied in their analysis is that foreign funders were not necessarily invited to play this role, but rather, given their own experience with research in the developed world, simply recognised the need.

Foreign funding for think tanks in an ideologically closed policy space, such as in Chile in the 1970s and 1980s or in Africa during the period post-independence, where authoritarian regimes developed, allowed think tanks to develop with relative freedom. State funding for alternative views was certainly not available, and it is questionable whether these centres would have been able to operate without the watchful eyes and sponsorship of the international community. Hence, State control can be challenged in this way.

A study by Matías Lardone and Marcos Roggero (Lardone and Roggero, 2011) in Latin America sheds some light on the way in which governments can affect think tanks via the funding they offer them. Lardone and Roggero surveyed 60 think tanks in the region and found that only 50% of them received any kind of public funding, and of these, some received as much as 66% and others as little as 5% of their overall budgets. In another study, Mercedes Botto (2011) found a large variability in the sources of funding in the region. In some cases, the percentage of think tanks receiving public funding was as high as 82% or 72% in Peru and Colombia, respectively, while in others as low as 13% and 24% in Bolivia and Paraguay, respectively.

Anecdotally, we know that the period during the first 2 decades of the 21st century saw an increase in the reliance of public funding for think tanks, fuelled by the rapid adoption of the evidence-informed narrative and methods like randomised control trials and systematic reviews as necessary tools for policymaking.

However, more important than the *amount* of funding received appears to be its type (Braml, 2004). Two types of public funding stand out in Latin America: programmatic and non-programmatic. The former refers to funds officially allocated through the national budget and that are usually focused on science and technology and awarded to universities (as is the case in Serbia and India, for example). Non-programmatic funding is more uncertain and awarded through short-term consultancy or opportunistic contracts, often dependent on personal links, networks, and the expertise or prestige of individual researchers or centres. Unlike programmatic funds, these are more readily available for economic and social research think tanks — and in the form of projects.

Programmatic funding is associated, across middle- and low-income countries, with government think tanks. Public programmatic funding is not exclusive: ZIPAR in Zambia falls under this category, but other donors and projects also fund it; the Kenya Institute for Public Policy Research and Analysis (KIPPRA) is an autonomous think tank which benefits from a number of funding sources, yet it was established under Parliament Act and receives funding directly from the public purse.

Programmatic funds also include long-term agreements between public bodies and think tanks, for example, to undertake evaluations of their policies and to provide ongoing assistance to parliamentarians, training to policymakers and new politicians, etc.

The nature of non-programmatic funding presents think tanks with serious challenges. To begin with, this kind of funding is piecemeal and thus limits their capacity to develop long-term, cohesive, and sustainable programmes. Funding uncertainty means that centres cannot hire staff, invest in data collection, embark on long-term studies, etc. Non-programmatic funding also responds to short-term political and technocratic concerns, and hence the contractual relationships that tend to prevail are more akin to consultancies. The high role of public funding reported by Peruvian think tanks in Botto's study reflects the growing demand for research consultancy, impact evaluations, and advice from the State, and not necessarily a commitment to support think tanks more generally.

The additional and undue power that the government wields over think tanks through non-programmatic funding came to light during the COVID-19 pandemic in Peru. The absence of criticism and contestation from think tanks, for instance, with respect to the government's education policies, is explained by the overreliance of think tanks on government contracts.

It can also be rather cumbersome depending on procurement policies, or closed off to new players if the adjudication or commissioning process is highly dependent on professional and social networks. Follow-up and incremental pieces of work are common in these cases.

On the other hand, non-programmatic funding can award think tanks more independence from the State. In fact, several of the think tanks I have talked to over the last year and that only receive this type of funding would not like to benefit from programmatic funds as this would seriously affect their independence — or at least the perception of independence.

State Intervention by Other Means

The statist view should not be reduced to a narrow focus on the direct monetary role of the State upon think tanks. In fact, governments can influence the formation and development of think tanks indirectly through a much broader set of policies and practices that have little to do with them specifically.

In South Asia, Srivastava found that immediately after independence, the State played a leading role in policymaking and in carving up controlled spaces for research (Srivastava, 2012). A strong State was partly

equated with a symbolic sense of independence — consultation would
have been seen as a sign of weakness — and the development of the civil
service conditioned the way in which research was conducted and the role
researchers could play. The State became all-pervasive in South Asia and
in the development of think tanks (Srivastava, 2012, pp. 7–8):

> *[The] development of think tanks in the early years of nation-building
> project was facilitated by the government itself which in most cases was
> geared towards generating data for policy documents or evaluating
> government's plans. This also made think tanks largely dependent on
> government funding, not to mention the fact that being 'close' to those
> in power which was the only way to ensure a better policy influence.*

Semi-democratic and even democratic states, more closely associated
with the pluralist perspective, can also exert an influence on think tanks by
defining the space in which they may operate. The legislation from which
the more traditional definition of the Anglo-American think tank is drawn
provides very clear boundaries of action that limit the think tanks' uni-
verse: They cannot be partisan, for instance, nor may they seek profitable
ventures. Civil society legislation is a common way of controlling their
activities, even if think tanks are not their main target. For example, in
Ethiopia, legislation has been used to ensure that foreign-funded local civil
society organisations, including think tanks, are not allowed to advocate
for governance and human rights reforms. In Ecuador, similar policies
limited civil society organisations' capacity to accumulate reserves and
engage more actively in policy spaces (Bellettini, 2011) and, in Indonesia,
not-for-profit organisations were made liable to income and value-added
taxes. Other policies such as tax legislation, employment, and immigration
policy, even, can have important effects on think tanks' capabilities.

India and Pakistan have used similar legislation to limit civil society
organisations' access to funding and allow the government to control and
even hold the power to approve or reject their projects. Peru passed similar legislation in 2025 that opens the door for organisations that are critical
of the government to be sanctioned.

Over the last 10 years, think tanks across the Western Balkans,
Eastern Europe, and the South Caucasus have been affected by a dramatic reversal of political freedoms. Multiple institutions of the state
have been captured, and think tanks, many of which emerged from the

human rights movements in the region, have been in the firing line (Stojanovic-Gajic, 2025).

In many cases, European Commission funds intended to support the development of civil society have been captured by the government, further restricting the sector.

In Indonesia, not-for-profit organisations were not allowed to bid for government contracts. But unsupported by a tax legislation that does not encourage philanthropy and in a context in which donors seek to work increasingly through the State rather than the vertical mechanisms that benefited think tanks in the past, some are finding their funding opportunities dangerously limited. Argentine think tanks have also been affected by unrelated legislation. A tax on all foreign transactions reduced the competitiveness of their researchers in the eyes of foreign funders as well as hampered their efforts to work with other organisations in the region.

Employment legislation can also affect think tanks' competitiveness and capacity to attract new staff. In Peru, new employment policies limit the time that students and recent graduates can work before the organisation must incorporate them into their payroll, with full employment benefits. Naturally, this also applies to think tanks. However, the nature of their work and the funding environment make it difficult for them to accommodate all these new legal requirements. Unfortunately, for labour rights expectations, think tanks must be flexible to grow and shrink as new issues are adopted and old ones dropped by researchers, policymakers, and funders. For young graduates looking for an entry into the world of research, policy, and even the corporate sector, the short- and medium-term projects that think tanks have historically offered them constitute an excellent opportunity — if offered responsibly.

Immigration policies also have the potential to strengthen or weaken the sector. In the past, I have argued that think tanks in countries with limited human resources, where their central banks and ministries of finance, donors, or the private sector have already snapped up the best researchers, should look further afield for their staff (Mendizabal, 2011a). By limiting the search to the local market, think tanks face having to pay crippling salaries and may, in fact, contribute to depleting the capacity of other crucial institutions for their own existence, such as universities and the civil service. A policy that facilitates the free movement of skilled labour into the private sector would have similarly positive effects on the think tank community.

There are other factors that affect think tanks in any given context. These could include policies aimed at controlling lobbying or political party funding and support, transport, and access to fast internet services.

The education policy is particularly relevant, too. The quality of the tertiary education sector has a direct effect on a think tank's capacity to recruit staff and source new ideas from full-time academic researchers. The subjects taught can affect the future development of think tanks' own areas of expertise; and the approach to teaching, for example, whether it promotes critical thinking or not, can dramatically impact think tanks' future effectiveness.

The International Development Industry

Where do donors come in? Are they part of the statist, elitist, or pluralist perspectives? Or are they simply foreign or de facto States?[1] In Latin America, Africa, and South Asia, donors have played a somewhat comple-mentary role to national States: When the public sector reduced its fund-ing of economic and social research in South Asia, donors stepped in (Srivastava, 2012). In Africa, the role of donors is so important that Kimenyi and Datta (2011) placed the power they exert at the same level as that of the State.

Throughout this book, I have included references to the influence that the international development industry has on the think tanks it funds. In a study of the way in which the British Department for International Development (now FCDO) uses evidence from research and evaluations, Harry Jones and I found that evidence was mainly used to make sense of ministerial policies rather than as the source of inspiration for new ones (Mendizabal and Jones, 2010). This policymaking model, in my own experience, tends to trickle down to the relationship between donors and the think tanks, research centres, consultancies, and non-governmental organisations, in donor countries and across the world, that develop and provide the evidence decision-makers need.

Think tanks are a tool — a source of useful evidence or advice — that international development funders use in pursuit of their own agendas.

---

[1]This is complicated by the emergence of private international development foundations which are increasingly the main source of funding for think tanks across middle- and lower-income countries.

Every think tank that works within this aid industry, willingly or not, is familiar with the term "policy-relevant research." Srivastava argues that this is the research that donors prefer: instrumental and empirical, rather than critical and theoretical (Srivastava, 2012). Jones and Young's (2007) review of research funding for DFID also worked under the assumption that this is the most desirable type of funding to be supported.

In the early 2010s, both Indonesia's SMERU and Zambia's CTPD, the former funded mostly by bilateral or multilateral donors and the latter by NGOs, saw a dramatic shift away from core funding towards contract funding. According to SMERU's director, this led to a deep frustration at the intellectual shackles the model imposed on his staff and at financial uncertainty, according to CTPD's head. In both cases, the organisations were forced to follow rather than create a demand for research on particular areas or issues. Their functions then became limited to legitimising or implementing donors' agendas instead of informing or influencing them.

Interestingly, although think tanks report that more funds are allocated to short-term contract projects, at least in public statements, funders appear to be shifting towards larger programmatic funding mechanisms. Why is this? Most of this funding is supporting global affairs, international development, or regionally focused think tanks rather than domestically focused organisations. The former are larger and therefore able to absorb more funding.

The Gates Foundation, another important and relatively new think tank funder, generally makes three- to five-year grants through formal agreements which include the expected results in the two main areas they cover: global development and global health. However, they frequently provide core grants to large intermediary partners, who then, in turn, provide funding and support to those undertaking research or providing services in the field. These grants can go from one month to six years of work, depending on the type of project.

As in the case of the FCDO, the Gates Foundation's support has been historically concerned with measuring the impact of its support, and this adds further constraints to what think tanks are able to do with the funds that they receive.

The Swedish International Development Agency (SIDA) also presents a picture of continuous core support to research centres in developing countries. This support, according to its own documentation, is directed at improving the centres as well as the general national research capacity in a number of priority countries. The Canadian International Development

Research Centre (IDRC) is also a well-known long-term supporter of think tanks' core activities. These were the leading partners in the Think Tank Initiative, which was originally devised by the Hewlett Foundation and counted on the additional support of the Gates Foundation, DFID, and the Netherlands Directorate-General for International Cooperation.

Multilateral donors like the World Bank and regional development banks are also important research funders in the Global South — crucially in Africa. In the histories of think tank formation and development, they are often associated with economic centres and Washington Consensus approaches and recommendations. The rise of think tanks in many developing countries is also closely linked to the application of structural adjustments and economic liberalisation programmes promoted by the banks. Think tanks in many cases legitimised and helped to implement these programmes by providing experts, advice, and public support. In the process, the relationships between them have developed and remain rather strong, and this makes it easier for researchers to access funds from them.

Nevertheless, even in light of this broad and diverse family of foreign funds, the experience of think tanks is that funding is becoming narrower, less concerned with fundamental questions and more with finding what works, and impatient about the slow adoption of new communication technologies and practices rather than the more urgent and immediate adoption. This can be explained by a number of factors: the demand to plan for and demonstrate policy influence, the role of intermediaries who impose consultancy-style and short-term contracts on think tanks, and the resurgence of the evidence-based mantra, among others.

Carving Out a Space within the State

This chapter complements the accounts of think tanks presented earlier. It confirms that think tanks are limited in their action by the agency of other players. In this case, the State, domestic or foreign, is conferred the leading role in the story. Think tanks then must work hard to carve out a space for themselves within the parameters set by the State.

This is in itself an interesting dynamic that adds nuance to the legitimising function of think tanks. The statist perspective awards the State the most prominent role among all other political players. Think tanks are thought of as part of tools of the State. This however does not necessarily imply that think tanks have no agency. In fact, the accounts described in this chapter and the cases referred to earlier show that theirs is a constant struggle and negotiation with the State.

The Chinese State may have the upper hand, but its political leaders, academics, and new entrepreneurs have contributed to shaping and reshaping the nature of think tanks and their relationship with the establishment for over half a century. Similarly, the funding mechanisms that affect and in essence control think tank development in Latin America or Southeast Asia are not simply accepted, but instead have led to the search for new business models and relationships to secure more reliable funding and work streams.

Inevitably, however, where the State is particularly strong, think tanks have had to learn to work with it and their models have adapted to serve it better. Not surprisingly, the 'legitimation' function appears to play a much stronger role from this perspective than from any other. Where the State has participated more actively — for instance, by seeking expert advice from think tanks — more academic think tanks have sought out additional funds from foreign or private sources to compensate or provide some room for manoeuvre. However, foreign funding does not necessarily imply independence: It can also come with strings attached. It is, after all, also provided by states (assuming most research funding comes from bilateral or multilateral funders, at least until recently).

Studying think tanks from this point of view confirms an issue that emerged from the cases in the previous chapter: *Democracy is not a necessary condition for the formation of think tanks*. They can emerge, as in the cases of Chile and Serbia, in response to the lack of intellectual freedom that authoritarian (and even totalitarian) regimes impose or, as in the case of China or Saudi Arabia, as the consequence of a public policy to strengthen the State's own policymaking capacity and/or to legitimate its own policies.

Obviously, think tanks can also contribute to democratisation processes. For this, they need to be supported appropriately.

Legitimation then is not the only strategy for think tanks in these contexts. Those that have been set up outside the State, particularly during periods of political and economic repression, by foreign initiatives or local civil society, have tended to fulfil more antagonistic functions in the little space available for them. This in turn has meant that they have become more dependent on the patronage of foreign donors and influential individuals or groups and have had as a consequence adopted a technocratic language, avoided ideological labels or advocacy practices, and have deepened their engagement with foreign donors or spaces to seek funding, validation, and protection. These strategies may not entirely counter the strength of the State but they do illustrate a more interesting relationship between the two. Certainly, this is not one in which think tanks are simple passive legitimators.

This perspective also demands that we pay greater attention to the nature of the State's demands and expectations of and from think tanks in our efforts to study them. Josef Braml's comparison of think tanks in Germany and the United States provides an interesting example of how to combine both an organisational and a statist perspective. On account of the German State's high level of influence in all aspects of German society, his analysis naturally awards a central role to the nature of the State. Similarly, it would be impossible to study Chinese think tanks without first reflecting on the characteristics of the Chinese State. The political science literature in Latin America that focuses on the politics of policies offers equally rich insights into the opportunities and barriers that exist for the formation, development, and functioning of think tanks (Scartascini, 2008; Stein and Tommasi, 2008; Spiller *et al.*, 2008; Mendizabal and Sample, 2009; Siavelis and Morgenstern, 2008).

Strong states are, therefore, able to promote strong think tanks. The Asian developmental states and the Chinese State examples have actively supported their formation and development. Weaker ones on the other hand can negatively affect them through unhelpful legislation and regulation. Greater institutionalisation, it seems, comes hand in hand with greater demand for knowledge in policymaking.

Chapter 8

Salomon's House: Intellectuals and Technocrats

A perspective that immediately draws one's attention in the literature awards elites a great deal of power over the policymaking process — and places think tanks within them either as members or tools of their political and economic power. Scholars like Abelson, Smith, and Ricci have written about this perspective (Abelson, 2006, 2009; Smith, 1991; Ricci, 1993). In this context, think tanks are described simultaneously as benign, scholarly, non-partisan, and genuinely interested in progress, as well as elitist organisations designed and oriented towards pursuing the interest of a powerful minority. The latter is a common view held among non-governmental organisations in some developing countries and some segments of the international development community.

Daniel Ricci's interpretation of this elitist perspective of policymaking is the *Baconian* metaphor of the search for Salomon's House — which relates to the platonic idea that it is possible to develop recommendations or blueprints for how an ideal society should be constructed and that these would come from a special group of citizens who have the capacity to discover the techniques and institutions that will keep society moving in the right direction (Ricci, 1993). This is also a story that social scientists (and many think tanks) aspire to and that is implicit in all the other metaphors that have influenced the formation and development of think tanks described earlier. It is also the story that underlines much of the account of the rise of think tanks across the developing world — driven by the individuals, philanthropists, the state, or foreign funders who shared the same idea.

The literature that describes think tanks as part of a cadre of elites is particularly interesting because it provides a window into the study of the roles of public intellectuals, technocrats, the media, and philanthropists — key protagonists in this story. It is particularly relevant in efforts to understand the lack of support for think tanks across the world: They are viewed as part of the elite.

Public intellectuals, central in Jeffrey Puryear's account of the rise of modern Chilean think tanks (1994) and the subject of a very relevant study by Richard Posner (2001), are credited with a role that is similar to that of many politically engaged (or ideologically identifiable in Braml's terms) think tanks: informing the educated public by undertaking analysis and commentary on matters of public interest. Technocrats on the other hand are perceived to be unaccountable to the public and focused only on the pursuit of designing and implementing evidence-based policy.

This distinction gained a certain profile in the aftermath of the Eurozone crisis, the Brexit process, and the rise of the MAGA movement in the United States.

In mid-2011, unelected technocrats charged with implementing the harsh budgetary cuts that the markets demanded, replaced the Greek and Italian Prime Ministers. In Europe at least, a public debate on the role of technocrats in democratic politics arose as a consequence of these events. *The Economist* published a series of articles on the role of technocrats and even questioned their legitimacy and value at the heart of democratic governments (The Economist, 2011):

> *Technocrats may be good at saying how much pain a country must endure, how to make its debt level sustainable or how to solve a financial crisis. But they are not so good at working out how pain is to be distributed, whether to raise taxes or cut spending on this or that group, and what the income-distribution effects of their policies are. Those are political questions, not technocratic ones. And they will not go away just because a technocrat has been made prime minister.*

The Economist articles set technocracies (which it likened to autocracies) against democracies, arguing that the latter are better at dealing with the decisions needed to address financial crises because they tend to be fairer — or are at least perceived as such. Democracies, and their politicians, just as public intellectuals, are freer to build their arguments by also appealing to non-technical values such as justice, fairness, duty, and even emotion. These provide a connection between evidence and action. And, since it is not necessary to be an expert to hold values, democracies offer greater opportunities for participation to the general public on issues of public interest than technocracies can allow.

Unfortunately, and particularly in times of crisis (and such is the permanent state of much of the developing world, at least as seen by the international development community), public intellectuals are easy targets for technocrats. Armed with powerful metaphors from the medical and physical worlds, they argue for the hegemony of evidence-based (and expert-based) approaches.

This distinction also resonates with the differences between the first generation of think tanks in the United States, funded by local businesses and individuals, and the second generation that depended on charitable foundations (Ricci, 1993), or between the ideologically identifiable think tanks and the non-ideologically identifiable ones described by Josef Braml (2004). These differences closely relate to the place that intellectuals and experts or technocrats hold in a society.

This chapter outlines some of the factors that explain the rise and fall of these two groups of interrelated players. Through it, I attempt to draw parallels between them and the realities and fates of think tanks. For the sake of argument, I allow myself to be inspired by the arguably "romantic" idea of an entirely intellectually independent and fully funded academic think tank, one reminiscent of Posner's independent pubic intellectual (Posner, 2001), and compare it with the more common dependent and contractual reality of think tanks that appears to be the norm in the developing world.

The media and philanthropists represent other political players that, from this perspective, play an equally important, if not more instrumental, role. In Chapter 6, we saw how addressing the politics of think tanks' origin and development sheds light on the importance that other institutions play. The media is one of them.

Equally important in the history of think tanks in the United States, as described in Chapter 4, is the role played by a number of domestic philanthropists, many of whom have appeared time and time again at the genesis of developing countries' think tank communities. In much of the Global

South, however, foreign funders rather than domestic ones have played a more prominent financial role, thus affecting the relationships between think tanks and other institutions (On Think Tanks, 2024).

Public Intellectuals: Knowledge and Politics

Public intellectuals may inhabit a world long lost to many think tank communities, but their journey is relevant to our understanding of these organisations and the pressures they face. The historical studies of the politics of think tanks provide some evidence that educated elites and public intellectuals in particular were able to play important roles during important moments of transition from closed to more open political and economic spaces.

The accounts by Jeffrey Puryear (1994) and Matías Cociña and Sergio Toro (2009) on the rise of think tanks in Chile have awarded public intellectuals a central role. In Africa, Kimenyi and Datta (2011) record a short-lived honeymoon period when, right after independence, public intellectuals enjoyed a constructive relationship with the newly independent governments. When the relationship turned sour, they went on to join academic and contract centres and non-governmental organisations.

Situated somewhere between academics and experts, public intellectuals are concerned with matters of the mind, of public interest, and of (or inflicted by) a political or ideological cast. They express themselves in a way that is accessible to the educated public with whom they engage directly and indirectly — through the media, for example (Posner, 2001). The public intellectual archetype is found in history books: Famous public intellectuals include Machiavelli, Milton, Locke, Voltaire, and Montesquieu, while more recent examples could include Noam Chomsky, Paul Krugman, Umberto Eco, Mario Vargas Llosa, Christopher Hitchens, Paul Collier, and Francis Fukuyama.

In today's social media-driven world, public intellectuals often step into the realm of celebrities with frequent television appearances, bestseller books on popular science and politics, and invitations to high-level advisory panels. Among other things, they help to popularise ideas (which are not always their own) and introduce the public to the academic and scientific method, something that is not necessarily well known or understood.

A key characteristic of Posner's romantic version of the public intellectual, and one that resonates with the idea of the ideal think tanks, is his or her independence — in this case, not being affiliated with a university or any other organisations. Of course, back in the days of the historical intellectual, the academic audience was not large enough to make good business sense, but with the appearance and development of the modern university — that hosts both teaching and research — intellectuals began to emerge from scholarship, as well. Noam Chomsky, Paul Krugman, Paul Collier, and Francis Fukuyama are all academics. Often, academic public intellectuals work primarily as academics and thus write for an academic audience first and foremost, while others have a more mixed approach, writing for both academic and non-academic publics, either directly or through intermediaries. Others, such as Umberto Eco, Mario Vargas Llosa, and the late Christopher Hitchens, are non-academics writing for both, or specifically for non-academics.

Today, non-academic or independent public intellectuals may be journalists or publishers, writers or artists, politicians or officials, influencers or celebrities with an agenda, and could even hold other "ordinary" jobs. The collapse in the cost of communicating first via social media and more recently Artificial Intelligence in support of content creation has created new opportunities for more to participate in several professional fields simultaneously.

It is not difficult to recognise the public intellectual's content: They communicate with intellectual superiority, are often careless with facts, and rash in predictions that tend towards the utopian and seek to steer society in a particular direction. When they comment on current affairs, they are opinionated, judgemental, and sometimes condescending, which was one of Hitchens's trademarks.

Ideally then, they should not be afraid of making risky logical leaps of the mind and going beyond the findings of known research to make value-based recommendations for action. As a consequence, they need different skillsets than those of experts. They must be able to develop and communicate compelling stories to a wide audience and exert authority through credence and persuade through rhetoric. More importantly, they need the institutional setting that allows them to practice their trade and develop these skills.

However, by the end of the 20th century, Posner argued that the number of independent intellectuals had shrunk and many who were able to support themselves through the publication of books or media appearances now had to resort to consulting or to more formal affiliations with universities and think tanks. This is also a common trend reported by think tanks in developing countries — whether that be in the form of think tanks that used to be supported through core funding, such as SMERU or CTPD, and that now are increasingly dependent on contract funding, or centres which have been set up as contract think tanks right from the start, such as CEPA.

What has driven this trend? Posner used an economic analysis to assess the demand and supply for public intellectuals. Their main product is not just information; they also offer entertainment, solidarity, and symbolic goods that provide a rallying point for like-minded people and institutions. This is very similar to the functions that think tanks fulfil, besides the production of policy ideas, such as providing and encouraging links between different policy actors or, as they did in Chile, by bringing together political parties and politicians from a formerly dispersed opposition (Puryear, 1994).

In the case of public intellectuals, demand for their goods comes mostly from the media — and particularly from editors. Public intellectuals have traditionally provided a service that stems from their capacity to apply a body of specialised academic knowledge to an issue to which it has not been applied before and that addresses a matter of current public interest or concern. That application, however general, requires certain intellectual expertise, and thus it is beyond the capacity of a single journalist or other communication professionals. The expansion of television channels, radios, online magazines, and other sources of digital communication, however, is changing the nature of this demand.

The advent of the web, 24-hour news, and the popularisation of the means of communication have led to an increasingly specialised media (Mendizabal and Yeo, 2010). Posner argues that this specialisation (and the media's focus on controversy) also forces intellectuals to specialise in their area of expertise. And, since the media pays attention to the credentials of those they call to comment upon current events, and they look for symbolic labels (think tank or university affiliations, recent publications, etc.), aspiring intellectuals will first have to take the long road towards becoming specialists. Posner worries that this prerequisite specialisation makes it difficult for intellectuals to pursue non-academic or non-affiliated careers. In his view, these affiliations can impose a number of conflicts of interest that hinder the public intellectual's work.

Think tanks are taking note of these changes, too. As Nick Scott has continuously argued for over a decade, think tanks are facing a digital disruption in the nature of the demand for their goods and services

(Scott, 2011). It is not just the media but also their funders and policymakers, think tanks' core audiences, who are changing the way they access knowledge and expertise. The compartmentalisation of professional and policy communities in the international development sector, which provides an important portion of policy research funding in the Global South, and the increasing complexity of policymaking at the global and local levels have led to more specialised demand for evidence and advice.

At the same time, there has been an increase in demand for a particular type of product from think tanks. For about a decade after 2010, the only evidence that mattered was evidence of *what works* — quantitative findings. Furthermore, in a context of increased public scrutiny on public budgets and policies, funders have been less willing to provide untied resources and have opted instead for issuing contracts tied to specific outputs and expected outcomes. There are some important exceptions, of course, but, all in all, these two factors have contributed towards reducing the space for more general reflection or the search for new problems, which are roles think tanks are in an advantageous position to fulfil.

It is possible to connect and compare Posner's analysis to changes that affect, positively and negatively, the supply of think tanks' "goods." First of all, the pool of public intellectuals is increasing as the media can now access intellectuals from other countries. The costs involved in the practice are also changing. Forced into academia, think tanks, or consultancies, the opportunity costs associated with being a public intellectual have rapidly increased. Once within a new organisation, intellectuals are faced with different sets of incentives — to publish, influence, and profit, respectively. Institutional affiliation can lead to other challenges. Universities, underfunded in many developing countries, cannot offer the space for intellectual exercise. In Zambia, for example, senior researchers in local universities commonly draw their income from consultancies and have, in some cases, resorted to joining government agencies and think tanks or civil society organisations, where the space for independent reflection and commentary can be even more limited.

Sometimes, however, affiliation can provide more space for intellectual reflection. In Peru, IEP is made up of an association of intellectuals and academics, some of whom have used the think tank as a vehicle to supplement their academic income and, aided by an important editorial fund, develop and maintain a minimum presence as public intellectuals.

The potential income, another component of the supply function for independent public intellectuals in developing countries, is also under pressure. Online access to free information limits what the media and the public are willing to pay for knowledge. Unable to match consultancies and nimbler think tanks, universities cannot retain "idle" scholars to reflect on matters of general public interest, and this greatly reduces academic intellectuals' pecuniary benefits. And, as we have seen, nor can intellectuals afford to remain independent for too long. Furthermore, in some developing countries where the pool of available researchers and consultants is limited by poor education systems and brain drain, the fees and salaries that the few competent ones are able to command can be significantly higher than any income originating from book sales, lectures, or media contributions.

The risk of making a fool of oneself, which Posner identifies as a necessary evil in the art of public commentary on issues and events that have not fully unfolded, also greatly limits what intellectuals associated with universities and think tanks can do and express. When I inquired into the reasons for why ZIPAR had not published more than a couple of papers in its first nine months in operation, I found, not surprisingly, that the think tank itself had been set up with several layers of quality controls, more appropriate for an academic institution than a think tank, which dramatically slowed down the process from idea to publication. But besides the formal process, researchers recognised that there were also psychological and political forces at play.

Not used to the idea of a think tank, most researchers new to the sector feel uncomfortable expressing their opinions on a particular issue without specific evidence to back them up — which greatly limits what issues they can offer advice and comment on. Equally important are the political (and eventual economic) consequences of expressing views on issues that may be both of public interest as well as of concern to the government. The nature of politics in Zambia, at least at the time before the 2011 presidential elections, meant that researchers who expressed their views in public (even if limited to recounting the findings of a study) risked being labelled by the media as either an opposition or a government sympathiser. I think that a similar degree of self-censorship took place in Peru between 2000 and 2017, when the elite bargained in favour of an economically liberal and socially progressive consensus (Mendizabal, 2022).

This reluctance to express one's views and opinions is found across think tank communities where the media and politics in general are deeply polarised — whether that be along ideological, partisan, economic, or social lines. To avoid confrontation, some think tanks resort to specialisation, adopting a depoliticised language, drawing their legitimacy from methodological expertise, and approaching their work and advice in a manner that turns them, in practice, into research consultants.

A recent trend, certainly across Africa, is the move from a local to a regional scope, seeking to avoid domestic politics, influence regional or global policy discussions, and capture incoming funding from funders eager to find intermediary organisations to work with. Invariably, this model further reduces the space for locally relevant public intellectuals to develop.

The dependence on contracts rather than core funding consolidates this. Contracts reduce think tanks' time frames and limit the room for manoeuvre that they would need to invest in long-term research and engagement initiatives. Their focus is thus only on incremental improvements to solutions to specific previously identified problems (Srivastava, 2012).

Conversely, it is possible to see how it is precisely the factors that reduce their space that also contribute to the importance of non-pecuniary benefits for public intellectuals: power, private satisfaction, and conviction. For both Posner and Puryear, the pursuit of a public intellectual role, even within an academic or policy research institution, is driven not so much by potential monetary gain but by the promise of agency, power, and recognition.

This account of public intellectuals is important because it mirrors many of the challenges that contract think tanks in developing countries face, but also because, in the absence of strong academic, media, and political institutions, think tanks have become a natural home for many public intellectuals.

Both Posner and Puryear conferred on think tanks the role of hosting these intellectuals and, in some cases, more than in others, providing them with the freedom and resources they require. Both were also rather emphatic on the positive role they can play in their societies — there is no doubt in their writing that they assume that a society that supports them is a more educated, more inclusive, more democratic one.

Technocrats: Political Naïfs?

Raucher (1978) called experts, or technocrats, service intellectuals, and Posner set them apart from public intellectuals by claiming that they are too focused on narrow fields of study — in a way suggesting that this is what public intellectuals become once they lose their independence. The parallel to think tanks is obvious to me: As more funding is channelled through contracts and demand changes from larger and more fundamental issues of public interest to a concern with solving specific problems or providing specialised advice, think tanks abandon their capacity to challenge the prevailing narratives and innovate, and must instead focus on serving and legitimating. Their functions change dramatically.

The work of experts is also different to that of public intellectuals in that the former tend to exercise their trade in private (Da Costa, 2011). And this has important consequences for the health of the polity.

Antonio Camou (2007) compared the role of experts in Argentine economic policy between 1985 and 2001. His study showed a significant increase in the influence of a smaller group of certain individuals and what he termed the privatisation of the institutional affiliations of these experts. The experts called upon to comment and provide advice on key economic policy decisions in 2001 were more likely to be affiliated with private think tanks, consultancies and international corporations, and private banks than had been the case in 1985. In fact, experts affiliated with political parties, 49% in 1985, were nowhere to be found in 2001.

Camou's reflection of the privatisation of policy echoes studies on the role of research contractors in the United States (Guttman and Willner, 1976). It is also particularly relevant to Ricci's historical account of the changing nature of the policy space in there (Ricci, 1993). Ricci's work reflects on how changes in national social values, among other factors, have contributed to increasingly complex and messier policymaking that has been accompanied by the privatisation of public policy debates. This privatisation or *technocratisation* of public policy was identified — and opposed — by Woodrow Wilson, who, at the beginning of the 20th century, saw the growing prominence of experts as a long-term threat to democratic institutions and as a potential impediment to full and open debate. He predicted that policy, if the trend continued, would only be discussed in new terms — jargon — controlled by a few (Smith, 1991). I think it is safe to say that the trend has continued even as, ironically, governments have privileged transferable managerial skills in which the concept of expert we commonly assume can no longer apply.

However, this process has not been linear. The history of the rise of think tanks in the United States shows how the driving forces behind the fundamental idea that think tanks could make positive contributions to policy had to be constantly renewed. The changes in metaphors illustrate this. Each time, the new idea responded to a faltering in the trust in science and an instinctive return to the more comfortable and inclusive world of ideological debate. In Latin America, Fowler made somewhat similar arguments in his study of ideology in politics (Fowler, 1997). In his view, ideology has become a pejorative term, synonymous with fanaticism. After the fall of the Soviet Union, political parties around the world suddenly renounced ideological discourses in favour of pragmatic ones (even if in fact ideology remained the main driver or force). This can explain the space that foreign funders had to promote local think tanks keen on economic liberalisation in the 1980s and 1990s in Latin America and Africa.

New Labour's Third Way, George Bush's Compassionate Conservatism, and the Conservative Party's "vote blue go green" slogan are examples of parties publicly accepting non-ideological (and contradictory) positions in the developed world. But even when the parties tried to change their ideological rhetoric, voters continued to see them through ideological lenses. After all, ideological discourse is a key element in the population's development of political thought, and its discourse is perceived as a definition of values that must be advocated by anyone who wants to represent the interests of a sector of society (Fowler, 1997).

Fowler also warned that when there is a vacuum of ideologies or of ideological debate, citizen participation can turn violent — nationalism and intolerance find no place to be explained and moderated by technocratic and pragmatic discourses, nor are they aggregated or merged with other views within a party or political community. It is then that the absence of ideological conflict can lead to nationalistic fighting and localism (Fowler, 1997). I would argue that the political apathy found among the main British parties today is in part responsible for the expressions of violence seen during the riots in the summer of 2011. Also, the violent resistance to austerity policies being implemented throughout Europe reflects a dissatisfaction with what seems to be a narrow technocratic solution to a problem that also demands a political settlement.

In Latin America, Fowler said, when the mood turns against political demagogy, political parties or their leaders inevitably abandon their ideologies in the name of pragmatism and technocratism, and yet, at the same time, they find the need to exploit or redefine them to justify their political existence. This paradox drives the recurring cycle between ideological and technocratic highs and lows described by Centeno and Silva (1998): When ideology fails to deliver and its misuse leads to political infighting, technocrats enter the political arena with promises of pragmatism and scientific solutions to social problems; but when experts fail to solve these prevailing problems and address the public's ideological needs, new generations of ideologues make an entrance.

Centeno and Silva went even further, agreeing with *The Economist's* argument that technocracy might even be undemocratic. Faith in this particular epistemological model of scientific policies — privileging what is taken to be an approach based on objective reason — denies the very essence of politics: the representation of particularistic interests and their resolution in some institutional arena. Of course, the idea that experts or technocrats are neutral to ideology is preposterous. After all, an ideology is nothing but a coherent system of ideas — a worldview. Technocrats as well as public intellectuals, and politicians, have them. And although they do not always attempt to hide them, they remain absent, at least in explicit form, from their arguments and appeals.

The international development industry is witnessing the resurrection of the faith in "hard" evidence illustrated by the popularity of impact evaluations and randomised control trials in everyday development discourse. It is a discourse that is in fact contrary and potentially threatening to that of participation and voice (Broadbent, 2012b; Da Costa, 2011).

The pressure on think tanks to specialise is hence being influenced not only by a change in funding arrangements but also by an increasingly technocratic discourse. The organisations I have worked with since 2011 in researching for this book and writing for On Think Tanks almost all agreed on the need to identify and employ researchers with quantitative research skills — particularly related to systematic reviews, impact evaluations, and randomised control trials. For many, this is more a resignation to what the context, particularly funding preferences, demands instead of conviction.

This external influence is particularly relevant because the rise of experts on certain policy issues and the adoption of certain ideological fundamentals is closely linked to the patronage of powerful international players and the role of a few universities in the formation of think tanks and policymaking professional cadres — for example, Ivy League universities in Latin American policy communities (Centeno and Silva, 1998; Camp, 1998). Across the developing world, spin-offs from Multilateral Development Banks or NGO projects have become standalone think tanks: SMERU and CTPD, for instance.

In more aid-dependent countries, technocratic policymaking may take a more perverse form. In a study on policy influence for the British Department for International Development (DFID), we found that policymaking on a number of issues of national great public interest (for example, user fees for health services, health insurance, and human resources for health) was taking place in private between donors and the relevant government officials — mainly at the secretary of state or ministerial level (Clarke *et al.*, 2009). This presents, in my view, an extreme case of Woodrow Wilson's prediction, where policy is not only being kept from the general public but even from local experts as well.

In my view, the rise of technocrats and contract think tanks may be accompanied by the fall or weakening of other political players, such as political parties as well as academic think tanks.

There is at least a persistent correlation between both. In fact, the story of the first foreign policy think tanks in the United States is about curtailing the powers of democratic politics and stemmed from the fear that foreign policy may be democratised and taken away from the hands of the "educated elite" (Raucher, 1978).

In Chile, however, think tanks developed and focused their attention on the promotion of public debates precisely at a time when politics was in the hands of technocrats. For two decades, across Latin America, several think tanks from Peru, Ecuador, Colombia, Chile, and Argentina, have implemented projects that seek to provide political parties with evidence-based advice during electoral periods. It is possible, therefore, to strengthen expertise while promoting political institutions.

The right balance

This is not an *either/or* reflection. The power of think tanks clearly relies on their capacity to occupy spaces of politics and expert knowledge. They can, in a very similar way as the romanticised public intellectual ideal could, apply technical knowledge to matters of public interest in a manner not commonly available to technocrats or politicians, whose institutional incentives limit their analytical and communication options.

It is important to consider, however, that if the agency of elites explains policy outcomes more than any other factors, think tanks and their staff and associates (whether they be public intellectuals or experts) can be expected to play an important role in policymaking, too. And just as elites are often criticised for preventing public participation, so could the role of think tanks be questioned: Are they also limiting the space and way in which policy is being made, or are they in fact contributing to making policymaking a more open and transparent affair?

The Media

The media plays a critical role in the elitist perspective. Finding and meeting the right balance between intellectual and technical practice greatly depends on the contribution that the media, as an institution, and its members, more specifically, can make. More importantly, the media can help think tanks fulfil their functions more effectively. Emma Broadbent's work on the role of research-based evidence in policy debates in Africa shows that two key determinants of research uptake are the existence of a public debate (often we do not know there is one) and the quality of the arguments being put forward (which are often unrelated to each other) (2012). However, throughout my interactions with think tanks' staff, the media is more than often presented as antagonistic or at the very least unhelpful to the promotion of a public debate. Consequently, efforts to engage with it have often met unfortunate ends: twisting of words, misrepresentation of facts, unduly editing of opinions to encourage controversy, etc. Expectedly, highly technical think tanks are more concerned than ideological ones. In Zambia, for instance, JCTR and PMRC had a clear narrative and were better equipped to deal with the media than ZIPAR. In Indonesia, SMERU was understandably concerned that it would lose control over its own research and communication process. In Peru, academic centres prefer to avoid institutional positions and insist that their researchers engage with the media in their own personal capacity.

Ideologically identifiable think tanks, on the other hand, tend to see the media not just as a means to a policy end but also as an end in itself. Guy Lodge, from the London-based Institute of Public Policy Research (IPPR), explained to me that in assessing IPPR's influence, he paid particular attention to the centre's media presence. The Peruvian Instituto Peruano de Economía (IPE), a think tank closely associated with the corporate sector, has similar concerns. Hans Rothgiesser, its former head of communications, once told me about a clear strategic effort to work with it and to position the centre's messages onto the media's public agenda.

CIPPEC, in Argentina, developed and implemented a very smart and aggressive strategy towards the media early in its formation. Right from the start, it sought to position itself as the go-to think tank for the media as a means to establish its credibility and reach a wider audience.

More generally, though, not enough has been done to understand the relationship between think tanks and the media, and in particular the factors that affect the way in which the latter uses the former and their research. In fact, even though studies such as Rich's and Abelson's are conclusive in that think tanks' visibility does not necessarily lead to policy influence, a great deal of attention is given to indicators that seek to measure it. This line of inquiry unfortunately takes a limited organisational perspective and treats the media as a passive audience, waiting to be influenced and used rather than as an agent of change in its own right.

Following this view, the growing community of researchers and practitioners focused on research communications, research uptake, or knowledge translation is packed with consultants, toolkits, and workshops focused on helping think tanks, researchers, and NGOs make better use of the media. Other initiatives, often involved in transparency and accountability, tend to treat the media as a development actor: a development NGO, in other words. Few pay sufficient attention to the unique nature of the media as a field, business, or sector, and the objectives and motivations of individual organisations.

Partly in response to this, the volume I edited with Norma Correa on the links between knowledge and politics in Latin America paid special attention to the relationship between think tanks and the media: Two papers, one by Ricardo Uceda from the Instituto de Prensa y Sociedad in Peru (Uceda, 2011), and the other by Pablo Livszyc and Natalia Romé from the Instituto para la Participación y el Desarrollo in Argentina (Livszyc and Romé, 2011), studied a series of cases from Peru, Colombia, Ecuador, Bolivia, Argentina, El Salvador, and Mexico.

The media has agency

Rather conclusively, the authors argue that a condition for a democratic society is an informed and educated public, and this is a role that the media can take on. In other words, the media plays a public role. But, as both studies also state, it must balance this public role with a private imperative: to make a profit. These often-conflicting objectives conspire to develop several possible strategies that help individual media organisations and outlets position themselves and secure a market and audience.

This suggests that between the fields of the media and think tanks described by Medvetz, there is a clear overlap of functions. In this case, it is the education of the public or, at the very least, the elites. The expression of this function, as well as others as we will see, is evidence of the media's own agency. Understanding this agency is crucial to understanding its relationship with think tanks.

David Hojman (1997) provides an excellent illustration of this agency. In a study of *Semana Económica*, an editorial section of Chile's *El Mercurio* newspaper, he attempted to answer some of the following questions relating to the paper's strategy during the early years of the Concertación government in the 1990s: What were the messages that Semana Económica was trying to put forward? Did the message change over time? What were the patterns of this change? How were they related to the acquisition of new information? Who were the messages addressed to, what were the messages' objectives, and were the messages successful in achieving their objectives? Without a doubt, these questions are equally relevant to the study of any think tank's communication strategy.

In developing the theoretical framework for his study, he explored several theories, including repeated cooperative games, informational lobbies, asymmetric information games, signalling games, and strategic information transmission. All of these analyse, albeit in different ways, the relations between the sender and the receiver of a strategic message. He assumed that even though the sender, Semana Económica, and the receivers, individual government officials, had different agendas, there had to be some overlap for there to be communication between them. In other words, his findings show both the sender and receiver had agency and how the former made use of the latter's.

Semana Económica had developed a consistent and simple message for the government, very much like ideologically identifiable think tanks are often able to do more easily than technocratic ones: "You are doing well, but you could be doing even better." More specifically, however, the publication advocated for specific policy decisions: a flexible labour market, lower taxes, an economy open to international trade, privatisations, the review of some of the government's measures against poverty and inequality, and the modernisation and liberalisation of Chile's financial markets and capital accounts, respectively. In other words, Semana Económica was advocating against changes to the policies that had been developed and implemented by the right-wing Pinochet regime and that it saw threatened by the new centre-left coalition. In this, it was not alone; other think tanks in Chile were working for the same objectives.

This is a situation typically faced by many think tanks in relation to their own audiences. Semana Económica's strategy is relevant to them, as well.

More importantly, Hojman's study shows that the media, in this case *El Mercurio*, had its own policy objectives, had developed a strategy, and used whatever means it had at its disposal to achieve them. When I asked about why few think tanks or researchers worked more closely with the media in Zambia, a related answer was offered: because the media, be it the then pro-government's *The Times of Zambia* or the opposition's *The Post*, had its own political agendas. In Indonesia, SMERU's reluctance to work with the media more formally responds to the same logic. By engaging with it, the centre, and its researchers, risks losing control over its research and messages.

In contexts where the reputation of researchers is closely linked to their ability to attract funds — as is the case of non-independent public intellectuals or contract think tanks — this is a risk that may be too high to pay.

For CIPPEC, in Argentina, the best defence against the media's own agenda and interests, has been the attack. Its heads of communications have run an extremely efficient media strategy, seeking out journalists and editors and maintaining a dynamic relationship with them. CIPPEC's leadership recognised this right from the moment of its foundation (Braun *et al.*, 2007). Its approach, however, demands that researchers adopt the organisation's brand and overarching policy messages. This is not always compatible with several of the academic or contract think tanks that have been described in this book.

This makes sense. The media's agency is not devoid of a discourse or narrative. El Mercurio had a message and so any evidence it sought to use was incorporated into it. The same is true elsewhere: *The Guardian's* online Development Portal, which was funded by the Gates Foundation, aimed to communicate a generally positive image of the development industry and showcase what the foundation itself calls "development progress." Its contributors, while at times critical, were by and large supportive of greater international cooperation.

Therefore, if the media has its own agency, and this is not surprising given that newspapers and broadcasters belong to larger media groups and corporations with public and private interests, how can think tanks safely engage with them?

If you cannot beat them, join them

Collaboration begins to be possible when we consider that think tanks and the media belong to the same economic or political elites and that they share functions, which they implement by slightly different means. Besides the elite education and policy-influencing functions already described, they share the roles of identifying policy problems and setting the agenda, providing oversight of public and private agents, and legitimising particular discourses or narratives. Instead of trying to beat them or use them, think tanks are learning to work with both the traditional and new media.

Media organisations demand news. According to Ricardo Uceda, news must be current (they must deal with issues of interest to the public today) and new (they should not be old stories, well known to the public, and predictable). Uceda interviewed journalists working in global and regional publications such as Michael Reid at *The Economist* and Tina Rosemberg at *The New York Times*. All coincided in their strategic use of think tanks to access relevant insights into news stories or, even, to identify which stories are worth writing about.

This is not restricted to dailies or weeklies. Bronwen Maddox, at the time *Prospect's* editor, told me that she and her team often enlist the help of think tank researchers to write articles for the publication. Think tanks are such an important resource for it that *Prospect* organised an annual think tanks award to recognise the contributions made by the most original and influential think tanks of the year on both domestic and international policies.

Bronwen Maddox went on to lead the Institute for Government first and then Chatham House.

Increasingly, social media has become a medium through which journalists look for think tanks and experts to inform them on a story or provide expert opinion. Think tanks as experts were, at least in the early years of social media, in high demand.

Less reactive are think tanks' attempts to develop long-term institutional relationships with particular publications or broadcasters, thus securing some degree of editorial independence. Uceda, for example, described how Colombian think tank directors have been historically more inclined towards the media and often publish regular op-eds. The origin of think tanks can explain this. Colombian think tanks, newspapers, and political parties share the same origin (Londoño, 2009). In Perú, Uceda and Norma Correa agree in their conclusion that think tanks associated with NGOs (or that have close ties to specific constituencies) are more media savvy and invest greater resources in engaging with it, compared to academic centres which are more comfortable communicating with peers or in private.

Uceda compared the Centro Peruano de Estudios Sociales (CEPES), a think tank that specialised in rural development, with GRADE and IEP in Peru. CEPES spent about one-fifth of its budget on communications. It published several trade magazines and ran its own radio show targeted at small-scale rural producers. GRADE and IEP on the other hand have made only modest investments in communications (although they have both significantly increased their attention to communications since) and focused their efforts on more traditional academic publications — particularly books, in the case of IEP. A similar distinction can be made with think tanks with corporate origins and associations. Norma Correa's account of the differences between academic, corporate, and NGO think tanks in Peru showed that both corporate and NGO think tanks spent more on communications and were more likely to have developed more sophisticated strategies and tactics.

... *Or replace them*

Increasing financial constraints on traditional media outlets are affecting their capacity to invest in research and analysis. This presents both a challenge and an excellent opportunity for think tanks. A weaker media field, just as weak political parties or governments, can affect the capacity of think tanks. As we have seen above, a strong media can not only help think tanks communicate their ideas but can also become a client for contract think tanks and a funder for more independent ones.

However, a weak media can also present opportunities for think tanks to fill this capacity gap. JCTR's aforementioned Basic Needs Basket and frequent op-eds by the centre's researchers provided ample material for *The Post* in Zambia. The survey that JCTR carried out could very well have been undertaken by a better-resourced national newspaper, which would then rely on experts for the analysis. In Zambia, as well, two publications, *The Bulleting and Record and Zambia Analysis*, responded to the lack of experienced and competent journalists by using research undertaken by national and international researchers and centres as well as inviting them to author the articles themselves. In Peru, the Consorcio de Investigación Económica y Social (CIES), a consortium of more than 40 think tanks, often organises training for journalists on key policy issues. They invite researchers to present their studies to journalists in a safe environment where both can discuss the nuances of the research, its findings, and implications. The same strategy has been pursued by centres in Zambia; for instance, Emma Broadbent reported that journalists participated in media training sessions run by USAID to improve the reporting on science, including the reporting on Genetically Modified Organisms in a balanced way (Broadbent, 2012c).

Global programmes like Sci.DevNet historically worked to develop the capacity of journalists in developing countries to report on scientific issues. This involves bringing them together with experts in developed and developing countries, and these experts are often based in think tanks. While these do not necessarily replace the media, they do fulfil some functions that could have been expected of newspapers or broadcasters.

Platforms like *The Conversation* have emerged to address this knowledge gap in traditional media and offer experts a friendly channel for their work.

Unfortunately, these are one-off and ad-hoc measures. And they are often associated with the promotion of a particular world view or policy and not so much with the capacity to develop and report stories. Where the media is seriously limited or disinterested, think tanks must resort to developing their own communication channels and, in some cases, compete head-to-head with traditional and new media. Think tanks' blogs, podcasts, web streaming of conferences and events, periodicals and magazines, partnerships with newspapers and broadcasters, and several other tactics provide them with more direct access to the general public. The ever-evolving digital space is making this possible at increasingly lower costs.

But, this goes both ways. It is increasingly easier for the media to generate evidence and play the role that think tanks ought to be playing.

Where Are the Philanthropists?

Philanthropists have played a central role in the history of modern Anglo-American think tanks. The account of think tanks in the United States provides several examples of how their roles have changed over time, and how these changes have affected the relationship of think tanks with their public and society in particular.

In the original think tanks, the local civil associations, philanthropists participated actively in the organisations. Money, agency, and the public came together within the centres. They were of course not the formal and professional entities they are today, and their capacity and reach were limited. Later on, philanthropic support itself became more profession-alised and separate from the centres' agency and public, and as a consequence so did the think tanks they supported. Larger foundations such as the Carnegie Corporation, the Rockefeller Foundation, and the Ford Foundation enabled the development of larger and far-reaching think tanks as well as a shift in accountability relations. Think tanks now owed themselves not only to the public and society they sought to help through their research and advice but also, and most importantly, to their funders. This separation is also palpable in the case of think tanks in developing countries, as we have seen in the cases described so far.

With a few exceptions, it is rare to find cases of domestic philanthropy directed towards research centres in low- and middle-income countries, certainly not social and economic policy think tanks (On Think Tanks, 2024). For the most part, ideologically non-identifiable think tanks, whether academic, contract, or advocacy, greatly depend on foreign funding.

It is not that economic elites are neither involved nor interested. But they do not yet appear to perceive the funding of research as their responsibility. While I was visiting SMERU in Indonesia, I met with one of its board members, a wealthy individual, whom I assumed would be an exception to the rule. But being a member of SMERU's board had not persuaded him that fighting poverty, and funding the research that helps do it, was also his responsibility. I have had similar conversations with board members in El Salvador, Ecuador, and Tunisia.

Support exists, but it is still sporadic and ad hoc.

Luckily, exceptions abound. Private sector foundations and individual philanthropists, including the TVS MOTOR GROUP, the Mahindra Group, Suzlon Energy, T.V. Mohandas Pai, and Rakesh Jhunjhunwala, founded India's Gateway House, a new foreign policy think tank. In Indonesia, private foundations and individuals also fund military and foreign policy think tanks. In Latin America, the Fundacion Getulio Vargas, in Brazil, is a funding powerhouse, supporting several economic, social policy, business, legal, and education research centres across the country. Argentine, Chilean, and Colombian think tanks, particularly those with clearer political or ideological affiliations, are also relatively well supported by domestic philanthropists. In Peru, the private sector has begun responding to the need to inform political debates: REDES was founded with an endowment from the National Association of Banks, and

Instituto Bicentenario is a new foundation that supports think tanks backed by Peru's largest Bank.

In general, though, the realm of public policy in parts of the Global South, has been captured by the international development community and from which the State has retreated (Srivastava, 2012) that lacks domestic involvement. As a consequence, think tanks tend to owe more to their foreign funders that to their domestic elites and public. The same detached relationship prevails here. Funders are large and foreign and the managerial aid agencies that deliver their funds through global or regional mechanisms have limited capacity to cater for the rich differences that exists between and within contexts.

What is the state of philanthropy in developing countries?

Philanthropy is not absent from developing countries, but it is not yet as well organised, and we know less about it, although it is certain that think tanks are not a priority. Erna Witoelar, chairman of the board of directors for the Asia Pacific Philanthropy Consortium, an association of philanthropic institutions dedicated to promoting philanthropy in the region, was quoted by *The Economist* as saying that the concept of philanthropy in Indonesia, as in many parts of Asia, is yet not well understood and that this results in low participation (Economist Intelligence Unit, 2011). The situation is similar in Latin America, where a Ford Foundation project also recognised that the very concept of philanthropy has a very different meaning in that region as compared to the United States.

It is possible to identify some general trends that may help identify how it relates to think tanks. Ambrose (2005) and a study by Fundacion ProHumana and Ford Foundation in Latin America (2003) have provided some insights that are worth reflecting on and provide a background to recent developments.

In general, there is concern and distrust among local philanthropists in low- and middle-income countries towards organised charity due to prior experiences with corruption and mismanagement within the public and civil society sectors. As a consequence, the Church continues to play a philanthropic role in the region, and it is perceived as a desirable channel through which to participate. It is no surprise, then, that the Church has played an important role in the formation and development of think tanks in Latin America and Africa.

Part of this distrust also stems from the absence of an institutional framework for the development of large-scale local philanthropic practice. There are not many adequate tax incentives and government support for philanthropic enterprises, thus limiting the establishment of large endowments and foundations. That said, corporate philanthropy has developed since the 1990s, and new foundations appear to be focused on the education sector, although social assistance, community development, the promotion of art and culture, health, and the environment are also important. Children and young people are the main beneficiaries of these initiatives, followed by the poor and specific vulnerable groups. Not many are dedicated to the promotion of human and civil rights or democracy. In practice, however, many of the new philanthropic initiatives do not have a redistributive social effect.

Unfortunately, many of these new philanthropic organisations do not have endowments or secured incomes and depend on their patrons for sustainability — and this ties them to the corporations or families that fund them. This, however, appears to be changing in Latin America and the Caribbean: Argentina, Brazil, Ecuador, Guatemala, Jamaica, and Mexico in particular are beginning to organise and promote more institutional philanthropy, according to the studies.

The situation in Asia is quite similar to that in Latin America. A certain degree of distrust of charities and NGOs, and tax incentives that are for the most part negligible if they even exist, does not help the development of philanthropic initiatives. Not surprisingly, as we have seen in the case of India, wealthy Asians often prefer to fund centres in the United States or Europe. The absence of organised philanthropy can also be explained by the existence of stronger familial and community safety nets that do not exist or are weaker in Anglo-American societies. In Asia (as well as in other developing regions), giving is more personal and local. According to a *Time Magazine* report, *Learning the Art of Giving* (Walsh, 2006), charitable donations in the region have traditionally tended to be a private affair, with the rich quietly giving directly to needy individuals within their family, religion, or village. Government corruption, particularly at the local level, discourages new forms of giving that may lead to less control and oversight.

This has, arguably, a positive effect. It means that traditional forms of giving tend to prioritise causes in which philanthropists can be involved. The relationship between the funder and the beneficiary is hence not as distant as one would expect from a more professional system. And this reflects the early relaticnship between funders and think tanks in the United States. *Forbes Magazine's* article, *When Asia's Millionaires Splurge, They Go Big* (Chen, 2007), gives further evidence of this process: Rich Asian philanthropists often take pride in their commitment to their causes. According to the article, those who contributed to charities devoted 12% of their wealth to charitable giving in 2006, the highest proportion among the world's wealthy philanthropists, compared with 8% for both North America and the Middle East, 5% in Europe, and 3% in Latin America.

When I spoke to Rohini Nilekani about her own decisions to fund think tanks in India, she expressed a very personal desire to promote informed public debate on an issue that was close to her interests (Mendizabal, 2016).

In Africa, the flow cf philanthropic resources is inspired by the prevailing view of the role of civil society: to fill the gaps left by the State and the market. But domestic philanthropy has to compete with the better-resourced and professional international development industry, which is able to capture the interest and resources of local NGOs. As donors shift towards larger and longer-term funding agreements with a smaller number of NGOs, however, new opportunities may appear, but this may demand additional investments. According to Ambrose, the scale of the African civil society sector remains small, constrained, and insufficient; limited by the amount of public sector and other foreign funding it receives in comparison to the unlimited needs of the population (Ambrose, 2005).

South African civil society has been more thoroughly studied (Kuljian, 2005; Ambrose, 2005). According to a study carried out by Johns Hopkins University (Ambrose, 2005), the South African non-profit sector receives most of its financial support from the government (42%), followed by self-generated income (34%), with the remaining 25% coming from philanthropic contributions from the private sector and international sources.

South African funders such as the DG Murray Trust or the Zenex Foundation are contributing to a growing number of African foundations, paying greater attention to research and think tanks.

Overall, then, the picture of organised domestic philanthropy that emerges is one that is still in its infancy but growing fast. In the process, it is being shaped by its own local values but influenced by global standards or models. A key characteristic of the emerging model is that in it, philanthropists remain closely involved in the causes they choose to fund and support.

But, if think tanks are not yet seen or understood as being part of the local social fabric, what chance do they have of being systematically supported by this growing sector?

Funding for think tanks

What proportion of this growing sector is focused on think tanks? Finding an answer to this question is nigh impossible. As we have seen, the term "think tank" is difficult to pin down to a single definition. Hence, what research does exist on the subject tends to include think tanks within the broader civil society sector. In fact, when discussing the role of the Church in the support of think tank in Latin America and elsewhere, it ought to be clarified that this is to a great extent due to its involvement with higher education more generally and that since many think tanks are hosted by universities or have been set up by part-time or former academics, their support could be seen as almost unintentional.

Secondly, organised philanthropy in low- and middle-income countries is a fairly new concept and is still developing its own institutional identity. In a way very similar to the think tank story, developing country philanthropists are copying and adapting the Western model of organised philanthropy. Consequently, to specify exactly how much is directed specifically to research and think tanks is still quite hard.

Thirdly, government control over civil society is still strong in developing countries and as a consequence few philanthropic enterprises choose a public route, preferring, instead, to remain private.

In any case, it is safe to assume that low levels of public investment in research provide a good indicator that think tanks are not at the top of philanthropists' agendas.

This may change. If developing country philanthropists want to remain closer to their grants, then they are likely to get involved in causes or enterprises that are close to their own professional and personal backgrounds. Education, therefore, tops the interests of philanthropists in developing countries and, although this has indirect effects on think tanks, it does not constitute direct funding. Orazio Bellettini, the founding director of Faro in Ecuador, considers that if there is an issue that private philanthropists are interested in funding, this issue is education. Not research on education policy: education itself.

The diaspora is also bound to play a role in shaping the funding pot and choice of causes or grantees. Diaspora remittances far outpace overseas development assistance. However, lack of information about local charitable groups, lack of trust in the process of overseas giving, and complicated legal processes are some of the challenges for deeper involvement in philanthropic giving (Sheth and Singhal, 2011). Nonetheless, since the diaspora is more familiar with the role that philanthropic enterprises have played in supporting think tanks in the West, it would not be surprising that their funds would eventually reach them. Recognising this, CIPPEC organised annual events in Miami and Madrid to tap into diaspora Argentineans' charitable interests during its early years. This complements a well-developed fundraising strategy that targets the few domestic philanthropists available, international development donors, and the national and local governments.

In general, philanthropic funding to think tanks is closely associated with education and ideological or partisan objectives. This latter category includes religious philanthropy. Indonesia has such a religious tradition of giving. Its Islamic religious obligations to donate to charity, like zakat, ensure that a majority of the population makes annual charitable gifts. Publicising one's giving, some believe, will undermine what is meant to be a selfless act (Economist Intelligence Unit, 2011). In Indonesia, education is also a top priority. Putera Sampoerna, an Indonesian billionaire, aims to permanently raise the level of the country's human capital through education, and so in 2006, he pledged US$150 million over the next decade to the Putera Sampoerna Foundation, which funds education projects.

Although historically Brazilian entrepreneurs have not engaged with civil society, a change came about from within by the emergence of an academic interest in philanthropy. In order to promote and expand this practice, Escola de Administração de Empresas de São Paulo, the business school that belongs to the Fundação Getulio Vargas, also one of the most prominent academic institutions in Brazil, set up the Centro de Estudos do Terceiro Setor (CETS) to promote the idea of philanthropy among the Brazilian society.

Research funding is also focused on the alleviation of more immediate suffering and the support of vulnerable groups. Health is another preferred issue among philanthropists. For example, the Fundação Maria Cecília Souto Vidigal (FMCVS) is a Brazilian family foundation dedicated to promoting early childhood development. They generate and disseminate knowledge to professionals who work with early childhood and also to managers who have a role in shaping public policies. The funds originally came from Gastão Eduardo de Bueno Vidigal, a banker, and his wife Maria Cecília Souto Vidigal, after the loss to leukaemia of the couple's thirteen-year-old daughter. Given the lack of research and knowledge about leukaemia in Brazil at the time and the need for investments in teaching, research, and care for patients, FMCSV was created with the intention of setting up, developing, and making available a haematology laboratory for the diagnosis of leukaemia. In collaboration with the haematology staff at the Medical College of São Paulo University (USP), the laboratory became a reference in research on haematology.

In a way, then, the involvement of philanthropists in developing countries follows similar characteristics to that in the early days of think tanks in the United States. Philanthropists want to participate in finding solutions to problems that are close to their own concerns — policy issues that they are familiar with or problems in their own communities. While service delivery is still seen as a priority, in countries where research is not prioritised by their governments, this should not come as a surprise.

Elites: Not a Zero-Sum Game

Viewed from an elitist perspective, think tanks belong to a group of society that has power over the fields of politics, the market, the media, and knowledge. They exist primarily for the purpose of educating or, at the very least, informing the elites who, by and large, make the decisions that govern nations and the world. Unfortunately, it also supports the idea of think tanks as legitimisers or enablers of elites' intentions. Depending on the view we take, this can be a rather comfortable or uncomfortable idea.

On the one hand, it fits well with the Anglo-American concept of think tanks and the historical progression witnessed in the United States, driven by prominent individuals, entrepreneurs, philanthropists, statesmen and women, and political leaders. It also fits well with accounts from low- and middle-income countries in which think tanks have been established by public intellectuals and academics, who are part of or at the mercy of the elites, experts, donors, and the State (and on which the previous section focused). With some exceptions, think tanks have emerged from the NGO sector, but this, too, corresponds to the educated minorities in many developing countries.

It fits as well with the anecdotal evidence of increasing overlaps between think tanks and the media, whether this is in the form of collaboration or competition.

On the other hand, this perspective directly challenges the *romantic* notion that think tanks speak truth to power, often heard in my interviews and conversations with think tank staff and funders. This idea assumes functions more closely associated with setting the agenda, auditing, and policy influence, and these are linked to the international development industry and its own agenda. Not surprisingly, NGOs in Zambia and much of the public around the world perceive think tanks as being part of the establishment: academics or experts working for the government, donors, or private interests, certainly not for the people.

It is also not surprising that Norma Correa did not find many Peruvian centres that labelled themselves as think tanks (before the launch of the Think Tank Initiative in 2010 in Latin America encouraged many "would-be grantees" to relabel) with the exception of those related, through funding or their business models, to the corporate sector.

More recently, this association with the elite has been used by populist regimes to discredit think tanks and experts — from the Brexit campaigners in the U.K. to the government of Victor Orbán in Hungary or Donald Trump's Republican Party in the United States.

Part of the blame for this reaction to the idea of think tanks as elites rests, in my view, with the roles hitherto played (or more accurately, not played) by the media and domestic philanthropy. The media, as the account of its role in creating and maintaining a market for independent public intellectuals shows, can contribute to opening the field of think tanks to the general public. In doing so, it could both affect their role as well as explain their value to society. Collaboration (but not collusion) can only help both: proving the media with invaluable knowledge that it needs to remain relevant and offer value to its audiences, and think tanks with exposure, feedback, and even income. It can also enhance think tanks' public education function.

Domestic philanthropists can play important roles, too. In low- and middle-income countries, domestic funding for think tanks can claw back their attention and focus from foreign to domestic concerns, and away from global funding interests. This could have the effect of bringing money, researchers, and the public closer together in a way not seen since the original think tanks in the United States in the late 1800s. And this can serve as well to support think tanks' elite education function.

Philanthropy in general has become increasingly partisan, aligning with progressive or conservative values or interests and the political parties that champion them — at home and abroad.

This alignment inevitably drags think tanks along — whether they like it or not.

Overall, think tanks can be easily described as elitist. Their presence is explained by the introduction, by elites, of a particular type of organisation and the pursuit of specific functions founded on exclusion rather than inclusion. This position, however, can be of benefit to the public good. The elites that think tanks are linked to are not homogenous and are instead represented by different institutions: the media, the church, philanthropists and civil society, political parties, etc. If think tanks can affect them, and if the elite are interested in their contribution, then they have the potential to influence to reach well beyond their own spheres of influence.

Chapter 9

Pluralism: The Marketplace of Ideas

The pluralist perspective does not view the State or elites alone as having much influence on policymaking — they can try to flex their muscles all they want, but in the end, many more forces take over. Pluralists view policymaking as the consequence of public and private negotiations between a large and mostly uncoordinated number of very different players, of which think tanks are just one more group among many. Their main role can therefore be seen as quite limited and indirect.

Pluralist perspectives appear in the works of several think tank scholars and focus on the policy networks that form around policy issues and debates, and it is within these that think tanks can be studied (Abelson, 2006, 2009; Braml, 2004; Smith, 1991; Ricci, 1993; Rich, 2006; Weaver, 1989). An excellent example is Emma Broadbent's work on the construction of political discourses around key policy debates in Sierra Leone, Ghana, Uganda, and Zambia. Rather than focusing on one particular actor, policy decision, or piece of research, her work "reconstructs" the policy debate by tracing the arguments of key actors and the "evidence" supporting each argument (Broadbent, 2012b).

The recognition that think tanks are just one more actor and that they are in fact much weaker than many others (Weidenbaum, 2009) is particularly refreshing. In a context in which international donors' monitoring and evaluation efforts are increasingly focusing on the influence that think tanks have on policy — and claims of influence are being made by anyone with any aspiration of attracting new research funding — this certainly helps put things into perspective.

Pluralism makes use of two of the powerful metaphors presented herein (Ricci, 1993): the Marketplace of Ideas, largely influenced by the incorporation of marketing practices into the world of think tanks, and the Great Conversation. These offer interesting alternatives to the more certain explanations provided by the previous perspectives.

Pluralism also demands a more nuanced approach to an important aspect of think tanks' objectives: influence. What is it that think tanks can influence in a context in which they play only a marginal role and no one in particular, not the elites or the State, is in charge? An elitist perspective would suggest that think tanks are influential if they affect the elites; statists would argue that the State must be monitored to assess think tanks' influence. But in this perspective, influence is much more intangible and even harder to track.

The pluralist perspective, as we shall see in this chapter, pays greater attention to changes in the way in which think tanks engage with their environment and the discourses or narratives that inform the public policy debate.

Democracy

Unsurprisingly, the aspiration for democratic progress tends to come hand in hand with pluralist perspectives. David Ricci's account of the rise of think tanks is focused on the breakdown of a controlled social order based on puritanical values in favour of a more plural — yet messier — society (1993). Democratisation processes are also partly credited with the development of think tanks in Latin America (Belletini, 2007), and as I have suggested, political liberalisation has played a role in Africa, South Asia, and Southeast and East Asia (Alvarado, 2024; Yeoh, 2024).

The effect that the transformation of societal values has had on the formation and rise of organisations similar to think tanks (and certainly the rise of the media, professional associations, interest groups, experts, political parties, etc.) is present in the literature on political intellectuals and experts in Latin America and as well as on Chinese think tanks (Zufeng, 2009; Shambaugh, 2002; Naughton, 2002; Glaser and Saunders, 2002; Gill and Mulvenon, 2002).

The recent attention to think tanks and the growth of the sector in Saudi Arabia and across the Middle East and North Africa since the Arab Spring further illustrate this relationship of political liberalisation (even if limited) with the emergence of think tanks.

Scholars who take this bigger-picture approach to understanding the relative role of think tanks pay a great deal of attention to long-term political, social, and economic trends. Ricci, for example, identified four trends that have shaped the development of think tanks in the United States (1993) that suggest an opening of the public space:

- The increase of experts being called to work with and from within policymaking bodies.
- An increasing dissonance in values within the political and economic classes (linked to the empowerment of minorities), and a rise of the consumer citizens.
- The adoption by leading think tanks of marketing ideas and tools to promote their work more widely.
- The increasing disorder in political institutions, as new ones are set up to address new challenges.

For Latin America, Orazio Belletini considered the effect that similar regional underlying trends have had on the formation of think tanks and their increasing importance (2007):

(1) Democratisation processes have opened new spaces and seen people from academia and civil society enter the government, thus creating new links with the "outside." Of course, the fact that Latin American countries have gone through several such processes (with democratic rule being followed by military regimes) means that this process has also been driven by cycles.

(2) More technical knowledge and information and communication technologies have made the job of the policymaker much more demanding; they must respond to demands from a more informed public, be more aware of their surroundings, consider more options, etc. Democratisation waves have also demanded highly qualified people to design and manage new institutions.

(3) At the end of the latest period of democratic consolidation (post-1980s), with several countries facing economic and political crises, the Latin American population and many international financing bodies lost trust in the State and its legitimacy was transferred to civil society organisations, including think tanks. This thus opened the policy space to new players, often better funded and staffed than the State itself.

Regarding Africa, Kimenyi and Datta (2011) stressed how political liberalisation and donor activity have affected the think tank community in a similar manner. The increase in non-governmental think tanks, they argue, could be explained as a response to the following:

(1) An increase in donor funding for and a perception of an expanding space for civil society.
(2) A prioritisation of policy issues related to political and economic liberalisation, as well as the "good governance" agenda, followed by poverty reduction and the Millennium Development Goals.
(3) An increase in funding to monitor and promote improvements in government policy implementation, which strengthened the position of civil society organisations as watchdogs or monitors.
(4) The role played by regional research funding bodies, such as the World Bank's African Capacity Building Foundation, which supported the establishment of think tanks across the region.

However, as we have seen before in the cases of Chile, Serbia, and China, political openness and democracy are not necessary factors for the development of think tanks. Instead, the idea that think tanks may contribute towards the development of democratic institutions is more plausible. The Chilean account and the recurring importance of think tanks in China as it opened and closed its political system offer evidence for this assertion. What emerges, though, is a particular type of think tank community, more intellectually diverse, different to that that which emerged in periods of authoritarian or military rule. Unfortunately, this difference is complicated by the messy overlap between economic and political liberalisation, which means that many new think tanks that emerged to promote a particular economic model may have been confused with efforts to promote political openness and plurality.

The Marketplace, Marketing, and Competition

Whether think tanks promote democratic institutions or democratic institutions promote think tanks, it is clear that what we see is increasingly open and messier politics. Pluralist metaphors like the marketplace and the great conversation can help us accept this complexity — that policymaking increasingly demands more experts and bureaucrats (within and outside of the state) to manage them, and that new political entrepreneurs from within and outside the system are becoming active participants.

The marketplace metaphor makes sense of this messier reality. Markets deal with disorganised and multifunctional agents and organise them into neat categories as sellers, buyers, and intermediaries. This language is prescient in many international development programmes and initiatives involving think tanks: Supply and demand side approaches, developing the knowledge marketplace, etc., are common phrases for anyone engaged with these initiatives. The concept of bridging research and policy also comes from this metaphor: The bridge is the intermediary between the sellers (researchers) and buyers (policymakers) of research.

The use of the term "broker," as in "knowledge broker," also stems from the marketplace metaphor.

Unfortunately, this approach does not always account for all think tanks' functions. While the traditional Northern literature on think tanks tends to see them as politically driven connecting or communicating organisations (Braml, 2004), international development practitioners and funders focused on the Global South tend to see them mostly as simple suppliers of evidence and advice, thus conferring on them the role of research producers. Think tanks are therefore awkwardly pooled together with universities, which, in more developed countries, would be mainly charged with different types of research (long-term academic work, for example).

As producers of knowledge, however, they are expected to fulfil rather significant feats: oversee the entire research and communication process with short-term funding. As a consequence, some think tanks in developing countries lose sight of the intermediary characteristics that are so well recognised and valued in the rest of the world. Instead, NGOs, networks, and the media, with or without the capacity to command research-based evidence, have appropriated this role.

Another challenge that this approach faces is that think tanks do not always operate in an actual marketplace. The buyers of the ideas are often not those who pay for them. Third parties, — NGOs, corporations, philanthropists, or donors — commonly provide the funding necessary for research. The audiences or publics that think tanks target and engage with may in fact not be at all interested in using their ideas. This idea, that the so-called users are willing buyers of ideas, is more relevant to consultancies than to think tanks — and so is the marketplace of ideas metaphor.

Nonetheless, this is a powerful idea. And it can, at its best, motivate think tanks to think beyond their preferred direct and private means of influence. It demands an understanding of the broader context, the market, and in particular, think tanks' clients or audiences.

The marketplace in particular offers an image of think tanks competing with each other for the attention and the business of a few clients or policy audiences. According to Ricci, the adoption of the marketing concept of the *audience as consumer* has replaced the *audience as the public* — the citizenry whose degree and depth of understanding of issues of public interest are crucial for the functioning of a healthy democracy (Uceda, 2011; Livszyc and Romé, 2011). Ricci's critique of the evolution of policymaking and think tank activity goes along these lines, too: By avoiding the public, think tanks may be contributing to undermining the public's right to participation in public life. Think tanks naturally tend to operate through private communication channels (Camou, 2007) or marketing of their arguments, targeting those with the power to make decisions, the elites or the State, very much as they would when selling a service or product in a niche market.

Boundary Workers: Connecting Vehicles

An alternative story that is slowly emerging from the literature in developed countries argues that in fact think tanks are in-between organisations, boundary players, and therefore fulfil critical articulating and value-adding roles in and between communities. This focuses on think tanks' functions and their relationship with others. Boundary workers, following Robert Hoppe's choice of term, can be found within the policy-making and research communities but also in a number of other societal fields such as the media, professional, and social networks, and business communities (Medvetz, 2008), with whom think tanks coexist.

Hoppe's definition of boundary workers positions think tanks as articulators of a fragmented space: central participants, interlocutors, or, in fact, the spaces in which conversations may take place — each with their own arguments and motivations. As boundary players, think tanks follow objectives and play functions that may also be part of the mandates of other types of organisations. Because of this overlap, both competition and collation are possible. The media, for example, often presented as the object of think tanks' communications and advocacy work, can also be an agent of policy analysis and political influence (Hojman, 1997), competing with think tanks for the attention of the same audiences.

Crucially, however, boundary workers must fulfil attributes that think tanks are potentially capable of (Hoppe, 2010):

- **Double participation:** People from the connecting communities should be represented in the boundary organisation.
- **Dual accountability:** The organisation or its members should be accountable to all the connecting communities and following their own rules. In other words, think tanks' staff should be able to participate and engage with politicians and academics with equal ease and confidence — even if this does not mean that they following their career paths.
- **Boundary objects:** They must produce content that serves the connecting communities such as econometric or climate models, biannual audits, reports, opinion pieces, and multimedia.
- Co-production is required rather than early consensus-seeking or compromise-building.
- **Meta-governance and capacity-building:** Think tanks must participate not only in the promotion of certain policy ideas but also in a certain type of policymaking and policy environment.

This perspective presents several opportunities to study think tanks as boundary players and find the spaces where they may be operating. It also helps to better understand the fundamental rather than empirical characteristics of successful think tanks if one considers them as such. But as a consequence of the focus on dual participation and accountability, it is possible to think of an alternative to competition.

Networks and collaboration

Collaboration can take a number of forms, but three that appear to be common among think tanks are: think tank networks, project partnerships, and think nets.

Think tank networks refer to associations of think tanks or their researchers around a particular thematic or policy issue. For example, the Latin America and the Caribbean Economic Association (LACEA) and the Southern Voice network bring together economic researchers and think tanks from across Latin America and the Global South, respectively.

National bodies are also relevant. The Economic Association of Zambia convenes economists and other interested parties from across the country to discuss and learn about economic policy. In Peru, the CIES is a network of policy research centres, independent not-for profit think tanks, and for-profit policy research bodies.

Other networks focus on policy spaces. For example, the European Think Tanks Group is another example of a geopolitically focused network of think tanks. It brings together leading European international development think tanks on specific European policy issues. The network aims to articulate a common message and amplify it to the right audiences across Europe.

More interesting are the attempts to develop a network of think tanks from BRIC countries, which include Fundação Getúlio Vargas (Brazil), Eco-Accord (Russia), CUTS International (India), Shanghai WTO Affairs Consultation Center (China), and the South African Institute of International Affairs (South Africa). This association targeted a geopolitical space and, in the same vein as Chinese think tanks, offers an opportunity for faster and more dynamic integration than what would be possible for bureaucracies to achieve.

In a similar vein, the Think 20 (T20) launched in Mexico by COMEXI during the Mexican G20 Presidency in 2012 and, later on, the T7 (in support of the G7 process) have sought to establish collaboration spaces among think tanks.

Other networks, such as the Atlas Network or the Stockholm Network, are known for their market-oriented perspectives. Rather than an opportunity or vehicle to work on a single issue, they convene their members around an ideology, which is in turn strengthened by their collective, but not necessarily coordinated, action.

While these networks can facilitate collective action, for the most part they offer think tanks and their researchers spaces in which to develop their knowledge on certain issues; access to national, regional, and global funding and policy spaces that they would otherwise not have access to; and, in some occasions, the credibility that is associated with the network and some of its members and that individual organisations or researchers may not have on their own. In some ways, these networks fulfil very similar functions to think tanks themselves. And this is why they join them. They work because, among other things, they focus on issues of core interest to their members (Mendizabal and Hearn, 2011).

Another kind of think tank network focuses not on a thematic issue, a policy space or process, or an ideology, but on the source of funding. The Think Tank Initiative, for instance, attempted to develop a network of think tanks it funds.

Many of the think tanks that made up the African cohort of the initiative used to be members of another IDRC-funded network, SISERA. After 10 years, the initiative's best legacy, in my view, is the formation of the Southern Voice network, which by 2025 boasted of 70 members across the Global South.

The Open Society Institute's Think Tank Fund also supported the establishment of a network of think tanks in Eastern Europe and the Western Balkans.

Unlike the earlier examples, these networks do not necessarily facilitate collective action around a common policy issue but could be more adequately described as coordinating mechanisms for the funding initiatives. The nature of their funding, however, does provide a certain degree of long-term stability in the networks that is not present in project partnerships. These networks also offer opportunities to facilitate access to additional resources and peer support.

Think tank networks, however, are hard to sustain. The network that formed informally among the grantees of the Think Tank Fund never developed into a formal alliance after the end of the funding. Efforts in Latin America, such as the Directores Ejecutivos de América Latina (DEAL) network or ILAIIP, whose members were the Think Tank Initiative's Latin American grantees, failed to take off. During the second phase of the Think Tank Initiative, several efforts were made to establish a network of think tanks in West Africa and South Asia. None succeeded.

Even the Southern Voice network struggles to ensure cohesion and collaboration between its members.

This is even harder at the local level. Think tanks tend to collaborate for very limited reasons: to access funding (e.g., as in CIES in Peru) or as mandated by a funder. In the 20 years I have been involved in this field, the relatively recent case of an open letter published by think tanks in Malaysia to sound the alarm on the government's response to COVID-19 strikes me as an outlier.

Think tanks are much better competitors than they are collaborators.

Project partnerships are more limited to specific time-bound initiatives. For example, DFID's Research Programme Consortia and large global initiatives such as the Climate and Development Knowledge Network brought together think tanks and other development actors to deliver very specific objectives.

The third type of collaborative relationship that is commonly found among think tanks refers to think nets. When the boundaries between organisations blur, so do the boundaries between organisations and their own resources — funds, knowledge, and staff. Stephen Yeo's work on think nets (Yeo and Portes, 2001) suggests that this is a possible future model for think tanks as they face increasingly regional or global challenges and as pressure on their own resources grows. The Centre for Economic Policy Research (CEPR) was the best example of this model. A think net based in London draws its researchers from universities across Europe. They come together on initiatives organised and facilitated by CEPR, usually at the European or global level, but CEPR also benefits from the work they undertake for their own universities and in their policy spaces.

In Peru, CIES often acts as a think net. The consortium has the capacity to organise research programmes drawing from researchers based in their member think tanks and so benefits from their own investments and reputations. Some of its members, recognised as think tanks in their own right, could also be described as think nets themselves. IEP, for example, draws the senior researchers from national and international universities. IEP, just as CEPR, benefits from these other organisations' investments in their staff.

ACET for Africa, a Ghana-based pan-African think tank, had a similar model in its inception. In both its research and policy advice, ACET relied on individuals from a network of African experts working independently or based in other organisations. This approach provided ACET with flexibility to mobilise resources when necessary, rapidly identify and develop new areas of work drawing from the expertise of their network experts; reach knowledge and policy spaces through their experts that they would not have been able to reach themselves; etc.

Another African think tank, the Africa Policy Research Institute (APRI), and REDES in Peru employ a similar model.

Joining networks, working in collaboration with others, and developing ones' own think nets involve important costs. Where some think tanks see the opportunity to learn from others and increase their capacity to respond to challenges, others perceive risks. The use of associates or consultants, in the opinion of SMERU's director, for example, presented a reputational risk for the organisation as it may not be able to control the quality of its work as well as it can control that of its full-time staff. Additionally, he considered that by working with researchers outside of the organisation, SMERU may fail to learn from its own work.

Working in project partnerships can be equally challenging. Depending on the type of initiative and the roles that other organisations play, think tanks may or may not be able to benefit from the opportunity. It is a widely documented complaint across the Global South that their researchers have seen their work limited to writing case studies or acting as translators for foreign researchers.

The same is true for think tanks' participation in networks. Effective participation demands the resources and commitment that many think tanks find difficult to command (Mendizabal and Hearn, 2011).

Relative independence

Networks involve complex relationships in which think tanks share objectives and resources with other organisations. This alternative to competition demands a more nuanced debate on the nature of "independence" — a key characteristic of traditional definitions of think tanks.

"Independence from" can be interpreted differently in different contexts. In Germany, independence from government funding is not so much an issue — what matters is the autonomy that think tanks have to choose the issues to study (Braml, 2004). The same could be said about British think tanks that are ideologically and financially associated with political parties, their leaders, or political leaders. For example, Ian Duncan Smith, the former leader of the Conservative Party in Britain, established the Centre for Social Justice in 2004 but secured an independent status by convening a non-partisan board (that includes Labour Members of Parliament) and a research agenda.

In the United States, on the other hand, where the private sector has been traditionally more trusted than the public sector in these matters, independence from the latter is more important (Weidenbaum, 2009) and independence from political parties is certainly crucial, as their legal status depends on it (Harvard Law Review, 2002). In China, it is in fact the patronage from important policymakers or prominent Chinese figures that constitutes a prerequisite for survival and the only way of ensuring a certain degree of independence in setting the think tank's research agenda (Zufeng, 2009).

The nature of international development and the dependence on foreign project funding in most low- and middle-income counties shift the discussion around independence to the relationship of think tanks with foreign rather than domestic funders. I have already discussed the way in which international development funding can limit the autonomy of think tanks to develop their own research agendas and influencing approaches. But it would be inaccurate to suggest an either/or situation. In reality, think tanks combine a number of business models, and the right combination can create opportunities that ensure their relative independence from any given source of funding.

Independence may also be interpreted to refer to the spaces that think tanks are free, or compelled, to occupy. Dependence on donor funding, for example, tends to be linked to regional — and even global — initiatives that aim to provide as much as one-size-fits-all support to think tanks as possible. Some national think tanks have become members or subcontractors of regional initiatives or networks and are often called on to engage in policy debates at levels, sometimes irrelevant to the policy issues they are working on. For example, AERC draws researchers from think tanks across Africa to discuss a range of economic policy issues, but there are not always equivalent bodies in each of the countries where their members come from, except for sub-regional networks or national umbrella associations. Other sector-specific initiatives (e.g., on health, education, climate change, and agriculture) further focus think tanks' attention to global or regional policy debates — sometimes drawing them away from more urgent matters. In one of her recent studies on the role of evidence in policy debates, Emma Broadbent argued that low interest from donors in Ghana's urban affairs was a key factor in explaining the weak role of research-based evidence in the debate over urban decongestion in Accra (Broadbent, 2012c). Other initiatives focus on specific geographic spaces (the Amazon Forest or the Nile Basin) or populations (pastoralists, women, etc.), somewhat brushing aside national and local politics and interests.

This discussion can also help to inform the elitist and statist perspectives. Affiliation or association with a dominant player does not automatically imply an absence of plurality. It is possible for dominant players to exert their influence on a policy space and, at the same time, enjoy a rather plural and diverse think tank community, as these will undoubtedly have formal and informal links to other players and communities.

A Greater Conversation: Supporting a Public Debate

The metaphor of a conversation is an alternative to the marketplace that focuses on the manner in which ideas and arguments are constructed in and shared with the public. While the conversation metaphor accepts that think tanks can compete with each other and other political players, it also welcomes the possibility of collaboration between them.

Broadbent's work on policy debates is a good recent example of this kind of approach. Implicit in her approach is the belief that a "good" public debate is desirable, and that something of the policy process itself and the contribution think tanks can make to society can be gleaned through its examination — a reflection of how "a society thinks and speaks" (Broadbent, 2012b).

Unlike approaches that start their analysis from the perspective of the researchers or a particular piece of research or an evaluation, or those that focus on a specific policy or policy change, hers did not assume any particular role for think tanks. Broadbent did not take an organisational, elitist, or statist perspective that may have influenced her inquiry. An organisational perspective would have focused her attention on the actions of particular think tanks; an elitist one would have focused the study on the relationship between think tanks and prominent individuals and groups, possibly including traditional leaders, donors, political leaders, and the private sector; and the statist perspective may have limited the study to the position of the government and its use of evidence.

This approach is hence more akin to that implied by the pluralist perspective in which organisations, elites, and States are part of a wider complex of social, political, and economic players. Some of these compete with each other while others collaborate along the lines described earlier.

By searching for the different policy discourses that are often present in a policy debate and investigating their development, it has been possible to break away from this practical path and explore others. Rather interesting findings have emerged out of the study, and they speak directly to the pluralist perspective and the role of think tanks as articulators of a wider debate — or a great conversation.

As expected, the functions that evidence-informed research and think tanks fulfil in these different policy debates do not fit neatly into one or another category. It is not only the instrumental function awarded by the organisational perspective, the legitimising function assumed by the elitist and statist perspectives, and the public education function conferred by the elitist view. Instead, they combine them.

So, how are these discourses developed and how could think tanks affect them and the outcomes of the public debate? According to Broadbent (2012b), the uptake of research by the policy debate is mainly affected by the following:

- The perception of the existence of a policy debate: The idea of a policy debate is a difficult one to grasp in contexts where policymaking is characterised by the dominant evidence-based policymaking discourse. The medical and efficiency metaphors that have driven the formation, development, and strategies of think tanks have affected, as we have seen, the roles they choose or are allowed to play in society. Unfortunately, these are often, by their own nature or the exercise of power by the elites or the State, limited to the fulfilment of technocratic functions in which debate is dismissed as political meddling and value- or interest-based policymaking.

  This discourse does not sit comfortably with public policy debates where arguments may draw their legitimacy and credibility from non-expert sources. As a consequence, funders, think tanks, and policymakers do not always present a policy debate to engage with and those who may have alternative discourses fail to see how they may be able to participate.

  It does, however, sit rather well with the agenda-setting and public education functions of think tanks discussed earlier. Rather than promoting a particular policy choice or course of action, think tanks can contribute to the conversation by helping the public identify and understand it. A good example of this was the Canadian Health Services Research Foundation's Mythbusters initiative that sought to challenge widely held beliefs about Canadian health issues.

- The subject of the policy issue being discussed: Some policy issues are likely to attract less attention from researchers than others. Research on genetically modified organisms in Zambia was supported by a global industry with clear interests in expanding their adoption and opposed by countless public and private organisations dedicated to preventing it. Naturally, there is therefore more research-informed evidence available to use or discard than, for example, on urban renewal and city centre decongestion.

Again, think tanks, besides focusing on influencing policy, can affect the policy debate by contributing to the generation of evidence on issues where little can be found. This idea of density, the amount and quality of information, and the means of communication available on a particular policy issue, first suggested by Peruvian journalist Mirko Lauer (Mendizabal, 2011c), provides a powerful alternative to the more linear descriptions of the link between research and policy. It is described in more detail in the following chapter.

- How like-for-like the different discourses or arguments are: The same forces at play in the imposition of technocratic policymaking approaches affect the possibility of like-for-like debates. The cases illustrate that often those people or organisations representing a policy discourse were unaware of the nuances of alternative ones. Those perceived as competing discourses were often not, and could very well be complementary (Broadbent, 2012b).

The media shone for its absence in the cases. In the discussion of think tanks as elites, we discussed the media's role in helping to strike a balance between ideological and evidence-based policymaking. Without it, it is also difficult to identify and participate in public policy debates.

Again, the role that think tanks can play turns towards intermediary functions, articulating the problems or issues being tackled, explaining or outlining the options available and proposed, weighing their costs and benefits, informing and educating elites and the public, etc.

- The locus of the debate: At the local level, within government, in parliament, in the court of public opinion, etc., the location of the debate is important because it can help to identify the relevant policy players and their discourses more clearly. Think tanks cannot be assumed to be relevant at all levels, and this will greatly depend on their context. Highly decentralised political systems as likely to offer opportunities to think tanks at the local level, but highly centralised ones are unlikely to. The executive and legislative branches of government can use evidence and think tanks differently, too.

- The involvement of international actors: Probably one of the least surprising yet interesting factors affecting research uptake is the role that international actors play in the formation of policy discourses and debates. Unaided by a domestic funding community, research bodies, including think tanks, often depend on foreign sources for funding;

unfortunately, this is limited to those policy issues in which they themselves are interested.

Besides funding for research, the adoption of ready-made discourses is also a common finding in the studies. Emma Broadbent found a rapid adoption of the language of international development that is reminiscent of the adoption of the evidence-based policy mantra and the Anglo-American think tank model. In topics as different as GMOs and chieftaincy reform, the roles of foreign agencies in funding the activities of local players and forming their discourses were equally strong and clear. Only in the case of the latter, however, did there appear to be the beginnings of a public debate — adopted, no doubt, from the GMO debate that has taken place elsewhere where these international players have met.

The importance of this factor resonates with the tension between domestic and foreign State power that Kemenyi and Datta put forward in their analysis of the political dimension of think tanks' formation and development in Sub-Saharan Africa.

It ought to be clear by now that this perspective is especially concerned with the existence and maintenance of plurality: intellectual, methodological, ideological, representative, etc. It does not limit itself to the pursuit of influence in itself as an objective, as it cannot accept that any one player or set of players has permanently more power over others and over the entire policy process.

Plurality works best as a way of beginning the study of think tanks in a particular context. It offers scholars the opportunity for a richer study of the entire policy community, including think tanks. And it provides think tanks themselves with a more honest and nuanced understanding of their role and specific functions.

Section 3

Why Think Tanks Matter

Chapter 10

What Then Explains Think Tanks' Success?

A Few Words of Warning

Assessing think tanks' success is not a straightforward matter. Are we concerned with think tanks' internal efficiency, their contribution to the policymaking apparatus, their influence over elites, or their contribution to a broader debate?

Efforts to assess think tanks' success tend to focus on their influence over policy and policymaking, arguably think tanks' most identifiable and expected function.

This is at the top of the list of concerns of think tank scholars and directors — and their funders. According to Andrew Rich, dollar for dollar, think tanks attract much more attention for their influence than any other organisation (Weidenbaum, 2009).

The literature, however, offers more than just a few words of caution in relation to focussing too much on assessing think tanks' influence. I think it is important to present some of the views of think tank scholars and practitioners, including my own, on the subject.

First, it is worth reminding ourselves that each of the perspectives described here offer different clues as to the nature of influence that is more relevant to think tanks. An organisational focus will no doubt concern itself with the strategic choices made by a single organisation and the influence it has had on a particular aspect of the policy process, as well as specific policies. An organisational perspective also places a great deal of emphasis on the think tank's own agency and treats other policy players as mere targets or observers of this agency.

The elitist and statist perspectives are likely to be more nuanced than this. The former would probably consider the effects that a think tank's research has had on particular social, political, and economic players, depending on the context. Influence on the media would be of interest, for example. The latter would pay greater attention to the State and how think tanks are able to function within it and the spaces that it makes available to them.

These, unlike the organisational perspective, would not treat think tank audiences as mere influencing targets but rather as active players with their own agency — and with considerably more power than think tanks.

A more pluralist approach, on the other hand, recognises the sheer complexity of the task. Broadbent's study presents an interesting account of the opportunities and constraints for think tanks' influence, albeit one that highlights the highly complex nature of the very idea of "influence" and how it sits with the concurrent objective of evidence-based policy. It is this perspective that informs the responses of think tank scholars and practitioners shared in the following section to the questions of what influence is and how it can be measured — if it should even be measured.

My favourite cautionary quote is from John Hamre, of the Center for Strategic and International Studies (Weidenbaum, 2009). In his opinion, as soon as a change happens, every think tank that had something to say about it would quickly claim it for itself. The proliferation of stories of change, a tool used by think tanks to demonstrate the impact they have had, is likely to come up with more than one such situation.

Unfortunately, there is no DNA test that can be administered to assess the undisputed paternity of a think tank over a policy process or its outcomes.

According to Nelson Poslky (Braml, 2004), one could only ask these types of questions if one ignored the complexity of the political process. While some causal relationships may be found in a few cases; for example, in contexts where think tanks are called in as consultants to solve specific problems or help implement a policy, any systematic explanations of this sort will remain an illusion. But even in cases where it is possible to show that a think tank's work has been used by a policymaker, it would be misleading to assume that this is evidence of its influence; is it not, in fact, the policymaker who chose to use its advice?

Specifically, asking these questions demands that we make a number of assumptions about policymaking and the roles that think tanks play in it, and that, in Robert Hoppe's view, do not hold up to scrutiny (Hoppe, 2010).

First of all, advocates for measuring influence expect to find an immediate knowledge uptake and implementation of advice that can only happen when a unidirectional transfer of advice is assumed to be true. Organisational, elitist, and statist perspectives that award think tanks very specific roles vis-à-vis other more powerful policy players may allow it, but the pluralist perspective fundamentally disagrees. This uni-directional transfer of advice in turn assumes that those being influenced do not hold their own beliefs, ideas, and agency, and are therefore inca-pable of influencing others themselves. In doing so, the object of think tanks' influence is in essence assumed to be incapable of choosing to be influenced or informed by one or more complementary or competing arguments. It denies them the capacity of synthesis, which is often at the core of policymaking — regardless of whether the policymaking process or its outcomes are deemed to be of good quality. In the face of it, even the elitist and statist perspectives place elites and governments in the driving seats of policymaking and so attempt to assess influence ought to recognise it.

This expectation also rests on the idea that policymakers and research-ers perform entirely different and exclusive tasks, and they therefore belong to non-overlapping communities. One only needs to review the histories of think tanks' formation and development described throughout this book to confirm that this is simply not the case. Think tanks, as well as public intellectuals and experts, co-exist with politicians, civil servants, non-governmental organisations, academics, the media, funders, and other political players. Their roles (given or chosen) sometimes set them to compete with each other, but cooperation is also possible and quite common. Their origins and missions are often shared and their staff move across boundaries with relative ease.

Given that collaboration, directly or via networks, is not uncommon, Kent Weaver (Braml, 2004) considered that it is hard to determine the policy influence of one particular think tank in relation to others — let alone all other actors involved. Not only is it difficult to draw clear bound-aries between them, but since they are all different, work on different aspects of a policy or with different policy players, seek influence in dif-ferent ways, there are many other players involved, and influence itself is also difficult to predict, it is difficult to target one's attention to a moment or process where influence may be accurately measured.

In other words, for the sake of measuring influence, we may miss out on all the other contributions that think tanks make to society more generally but that are impossible to measure.

Can we rely on reputation for an assessment of success? According to Leslie Gelb, of *The New York Times*, think tanks' influence is so episodic and arbitrary (Abelson, 2009), meaning that past performance cannot necessarily predict future returns and relying on reputation to rate or assess think tanks can be misleading. Echoing this, Abelson and Ricci have warned that as policy communities become more open and complex, as is the case in developing countries undergoing economic and political liberalisation, it also becomes more difficult to say anything meaningful about the contribution of specific think tanks to those communities. Hence, they may be well known but that does not necessarily mean they are making substantive contributions to policy and policymaking.

As a consequence of this undeniable complexity, Hoppe concludes that a robust inquiry into research uptake and policy impact would be so costly and time-consuming and yet never able to address the attribution problem, that nobody would be able to afford it (2010). A manageable set of indicators focusing on that which can be measured would only lead to conformance and perverse conduct. Think tanks would, as discussed earlier, lean towards private means of influence that could be "controlled" rather than the promotion of public debate and the involvement of other societal actors. Furthermore, the lessons drawn from these studies would still be anecdotal and irrelevant for all other situations.

Not surprisingly, then, when asked what approach would best address this question of influence, David Frum, formerly at the American Enterprise Institute, said that since one cannot measure influence, none would do it (Weidenbaum, 2009). This should lead us to interpret many think tanks' resistance to evaluating their influence not as stubbornness but rather as intellectual incredulity towards the credibility of the approaches offered.

Finally, even enquiring about influence may not be a good idea (Hoppe, 2010) — and influence itself, at least when narrowly understood as bringing about policy change, may not be the only desired outcome. If a think tank is so influential that all the policy proposals that it makes are always adopted by a government, we could safely assume that either the government is particularly weak and easy to influence or that the think tank is too strong for the sake of democratic debate, or that the think tank is acting as a *legitimator* of government policy or simply behaving as a consultant to its predefined policies. While this may resonate with the interests of many development donors, it may not be compatible with the principles of institution-building that many also advocate for.

So even if we were able to describe the relative influence of a particular think tank on a policy or policymaking, making this influence explicit may be counterproductive for the policymaking space and for some think tanks in particular.

Take, for the sake of argument, a think tank in a country like Ecuador, which receives all of its funding from a few bilateral donors and some international think tanks and NGOs. As the Ecuadorian government of Rafael Correa became increasingly suspicious and wary of the influence that certain foreign players had on domestic politics, it became risky to advertise policy influence. In fact, as many think tanks in Ecuador, Venezuela, Nicaragua, and El Salvador have found out over the last decade, even highlighting their relationship with certain funders can be dangerous. The same concern is voiced by think tank leaders across the world. Nationalist and populist discourse is challenging think tanks' hitherto efforts to be transparent and open about their funding sources. Association with certain global foundations or bilateral agencies is fuelling claims of undue influence and meddling (Lipton *et al.*, 2014; Mendizabal, 2025).

Had Chilean opposition think tanks boasted of their influence and capacity to convene politicians during the years in which they were successfully supporting the development of opposition public intellectuals, the story would have turned out quite differently.

When I visited ACET for Africa in Ghana in 2011, its staff described a similar challenge for the think tank. Its value proposition rests in its capacity to work directly with politicians and policymakers at the highest level of government — advising presidents, prime ministers, and cabinet-level ministers. If they were to boast about the influence on and the roles they played in specific policies and decisions, even if they were in fact responsible for writing or implementing them, they would soon lose much of the access they enjoy today. ACET's staff recognised that while they may provide the expertise, the decision to use it is a political one and not theirs to make.

The view that think tanks can and should measure their influence (and report on it) also assumes that they are independent from those that they

214 *Exploring Think Tanks: Diverse Origins, Evolving Futures*

seek to influence. It follows the romantic idea that they speak truth to power without consequences, one that may be possible in Posner's ideal account of independent public intellectuals or in a utopian pluralist democracy where ample domestic funding allows think tanks to think and influence independently. Unfortunately, the fact is that many think tanks work in contract mode and with short- to medium-term funding horizons that constrain what they can and cannot do. And they are more likely to owe their dues to an elite and/or the State.

Furthermore, as we have discussed in previous chapters, most think tanks exist in a space that is not free from the limits imposed by other, more powerful, actors. They can speak truth to power, but in the terms that power allows.

As for policy change being the sole objective of think tanks, again, the literature provides alternative accounts. Throughout their history, think tanks have not only contributed to changing policies but also affected the policy environment — promoting the powerful evidence-based mantra, for example, that informs how policymaking is conducted. As we sought to describe think tanks by their policy audiences or influencing targets, this idea was briefly discussed. Think tanks may attempt to influence more than just policy outcomes: Individuals, processes, and the environment in which policy is made are equally important. Even more importantly, think tanks may seek to prevent change rather than promote it. The think tanks that emerged in Chile in the 1990s, in a similar vein to publications like *Semana Económica*, sought to maintain the status quo (Hojman, 1997; Cociña and Toro, 2009).

My conversations with think tank leaders over the last 20 years are matched by discussions with entry-level staff. Why join a think tank? Absolutely, the most common answer is to develop skills necessary for their future careers. One of a think tank's core functions is to develop and train future cadres of policymakers.

Finally, I think the focus on influence biases our analysis. In a recent reflection, I suggested that it is not influence but "usefulness" that we should be concerned about (Mendizabal, 2024).

> ... *the fixation on think tanks' influence is misguided. Their true value lies in their ability to provide useful contributions to their political systems. By shifting our focus from influence to usefulness, we can better appreciate and support the vital roles that think tanks play. Think tank leaders and funders should embrace this perspective, recognising that*

their greatest impact comes not from shaping policy directly but from supporting the broader political ecosystem.

This long introduction outlines some of the factors that make any meaningful evaluation or assessment of a think tank's influence a seemingly near-impossible task. A task that, all too often, think tanks and their funders focus too much on.

So, What Explains Think Tanks' Influence?

Despite these challenges, the literature is peppered with attempts to assess think tanks' influence. Although rarely recognised, these are influenced by the perspectives that have been outlined throughout the book. As a consequence, they offer somewhat alternative accounts of the factors that explain think tanks' influence. Brought together, however, they begin to paint an already familiar image.

Emma Broadbent's study offers a starting point for this exercise. Her work confirms some lessons already identified by the literature. We know, for example, that evidence is more likely to influence policy if there is sufficient evidence available, in other words, high density (Mendizabal, 2011c). She also found that both the length of the policy process and the spaces in which it unfolds matter (Rich, 2006). These are relevant to the study of the influence of think tanks, too. Where more evidence is available, think tanks may be more influential if they can help make sense of it, or they could fill in gaps where this is not enough. The longer the process may offer think tanks greater opportunities to participate and, depending on the leading institutions, fulfil different functions.

Visibility Helps But It Is Not Enough

When assessing influence, most studies tend to focus on a research programme or initiative or the policy change they sought. They attempt to track, using episode or case studies, the causal chain of events from inputs to outcomes: from research output to policy outcome. These are sometimes referred to as theories of change. More nuanced studies accept different interpretations of policy and therefore consider changes in discourses, attitudes, and behaviours of policy players as well as changes in procedures and the content of policy (budgets, legislation, etc.)

(Jones, 2011). The focus of these studies, as one would expect from an organisational perspective, is still on the actions of a particular think tank (or other research producers or intermediaries), a single story or pathway of change, and an identifiable and ideally measurable change.

This is in essence the application of an approach more appropriate for ex-ante planning to an ex-post assessment or evaluation of change. However, when this is adopted for evaluation purposes, the linearity is no longer useful. Why should this particular strategy be the focus of inquiry and not any other unexpected effects that think tanks' activities may have on their immediate and more distant environments? And what does it matter that a think tank was successful in a particular instance if it is not in all other cases?

In order to address this, other studies, such as the global and regional rankings produced by James McGann, focused on think tanks' visibility or popularity as a proxy of their overall influence. More nuanced assessments of single think tanks attempt to map out their presence in the media or the Internet. Indicators such as web hits, the number of times their work has been shared via social networks, the number of followers, and other such measures are used to assess their visibility. Audience or policy surveys like the ones used by the Think Tank Initiative can also be employed. These surveys ask potential audiences about individual think tanks or pieces of research. They seek to assess whether they are known, if their work is relevant and useful, etc.

Unlike more qualitative stories of change studies, these do not focus on the influence of a particular piece of research or on a policy change. Instead, they assume that a think tank is bound to be influential if it is perceived to be so by its peers and direct audiences. This link, however, demands a significant leap of faith.

There is certainly a link between these two variables, visibility and influence, but it is not a straightforward one. Several studies of think tanks in the United States have shown that visibility can help, but it does not necessarily lead to influence (Abelson, 2006, 2009; Rich, 2006; Weaver, 1989). In developing countries, anecdotal evidence shows that the same is true. Two perfect examples are SMERU and CIUP. None are particularly visible in their own countries beyond the policy communities they inhabit and have not made, at least until now, much effort to significantly change this. Nonetheless, their influence on policy substance is relatively high — and obvious for anyone who cares to inquire about it. There are a number of reasons for this.

First, as Andrew Rich has pointed out, visibility is often associated with media presence, which tends to take the form of commentary or opinion rather than proposals and estimates (Rich, 2006). The same is true for an organisation's online presence. Traditional and new media coverage may simply reflect the final stages of the policy process when think tanks, as well as others, are being called upon to comment on a decision or policy that has been made or implemented. Social media presence may simply reflect the public's awareness of the issue and its demand for clarifying information.

Participating in the policy process does not always denote influence. In some circumstances, visibility regarding a particular issue can lead to think tanks being invited to participate early on in the policy process (Rich, 2006). Well-known think tanks and their experts will be in a more advantageous position than their less-known peers. But if the think tanks are invited to provide evidence to justify or support a policy discourse or decision, a situation that is recognised in the literature and by think tanks across the world, then one could argue that influence is more likely to have taken place in the opposite direction.

Andrea Baertl's work on think tanks' credibility adds weight to this discussion. Credibility is highly influenced by proxies such as visibility and membership to certain networks — it has little to do with the quality or robustness of their research or recommendations (Baertl, 2018).

Ideological Affiliation — Or at Least the Perception of One — Matters

A second explanation is that think tanks are not neutral. The fear held by many think tanks that the media may unexpectedly brand them with their own ideological biases is well founded, but not because of some evil media plot to undermine them: Think tanks, like all other actors interested in affecting policy, are inherently ideological. Even the rejection of partisan politics and the wholehearted adoption of the evidence-informed mantra are ideological choices in themselves. Their funders, the researchers who work for the think tanks, and the think tanks' own networks and audiences play a part in making them so.

For example, Rafael Loayza (2011) found, in the case of Bolivia, how different development agencies had funded two groups of think tanks: The World Bank and the Inter-American Development Bank had provided funding to think tanks closely associated with the Washington Consensus and economic liberalisation, while European bilateral donors and NGOs had mainly supported think tanks closely associated with the anti-globalisation movement in Bolivia and rights-based groups. The latter were in turn associated with Evo Morales's Movimiento al Socialismo (MAS) political movement and played an important role in his ascending to power and the policies his government pursued. Researchers themselves also join think tanks with some understanding of what the institutions stand for. Think tanks' own associations, as we have seen before, also indicate their ideological biases.

These perceptions matter and determine who has access to the relevant policy spaces and who does not, and therefore it is important to know who leads the policy process. In some contexts, ministers will have the capacity to demand and use evidence more than parliaments, for example. In other cases, opportunities may arise from an issue being adopted by the media and driven by it. Having access, through affiliation or affinity to them and their ideas, would go a long way in explaining think tanks' influence.

Therefore, think tanks' ideological alignment with the government or with other powerful actors can help explain — even predict — influence. The Heritage Foundation was expected to be influential in Donald Trump's second government; nobody would have expected the Brookings Institution to be influential.

Setting the Agenda and Developing Policy Options

Other studies attempt to unpack these associations and the specific circumstances that explain their influence on different groups of policy players and their discourses. They recognise that the measure of influence cannot be limited to particular research projects or policies and must instead address a broader set of possible contributions that think tanks can make to society (e.g., Broadbent, 2012a, Braml, 2004, Mendizabal, 2024).

As mentioned before, visibility does not necessarily lead to substantive influence. This is more likely to happen if think tanks were involved in setting out the agenda (identifying the problem or framing the discourse) and providing well-thought-out and developed policy options and estimates (Rich, 2006). Think tanks that limit their role to providing problem-solving advice, commentary, and short-term analysis only when there is demand for it often find that they have access, but this does not translate into substantive influence. Instead, it is limited to operational support and maybe a contribution to public dissemination.

This involvement, in turn, is heavily affected by the real or perceived identifiable ideology that conditions think tanks' access to certain private and public policy spaces and the use of the most advantageous communication channels. Therefore, the capacity of think tanks to combine research-based evidence with appeals to values and other sources of influence (political and private interests, legislation, tradition, pragmatism, etc.) in developing their arguments, for instance, as JCTR has managed to do, is instrumental in explaining the influence they may have on their environment.

Influence Depends on the Degree of Institutionalisation of Others

The opportunities for think tanks to influence policy or to make a contribution to society are often beyond the reach of their own agency and in the hands of others. The degree and type of institutionalisation of other players — the State, political parties, civil society, the market, the media, academia, etc. — play a critical role in think tanks' capacity to influence. The review of both the northern and southern literature points in this direction: Strong states and political parties (democratic or not) create opportunities for think tanks to participate in the policy process (Mendizabal and Sample, 2009; Correa and Mendizabal, 2011). This participation may be in a context of political plurality in which think tanks are freer to seek to set the agenda or are limited by a dominant policy discourse and the space and functions provided by the elites and the State.

Braml has delved into this issue in great detail. His comparative study between the United States and Germany has identified a number of factors that explain the apparent proliferation of think tanks in the former and their different nature and style of influence in both. But in both cases, the degree of institutionalisation of the political system is critical in explaining think tanks' strength. Chinese think tanks have also benefited from a strong State that has recognised their importance — even if it has, at times, been rather unfriendly towards them.

The media too, partisan or not, will be more likely to use think tanks if it is well funded, staffed, and managed than if it is not. A media sector with clear roles and objectives — even with an ideologically identifiable editorial line — is more receptive to using think tanks, their experts, and their evidence. And it is also likely to support think tanks in more public functions, thus supporting them in the delivery of their mission. A strong academic sector also guarantees think tanks' most valuable asset — researchers — and offers them intellectual support by focusing on more fundamental and less immediate research questions. The private sector is another important purchaser and user of research and analysis. It is also a source of philanthropic support indispensable for think tank autonomy and financial sustainability.

The issue then is not only whether think tanks alone have the capacity to influence; do they have the best researchers, great communicators, lots of funding, etc.? More important appears to be the degree of institutionalisation and the role that evidence-informed arguments are given by these other sets of organisations. Vietnam and China may have single-party States, but research-based evidence plays an important role in both the Communist Party and the State; consequently, links between evidence and policy and politics have developed and supported the formation of think tanks. In the United States and in Britain, political parties and the State are dramatically different. The former has relatively porous parties and a presidential system, while the latter has closely managed parties and a parliamentary democracy. Each offers different kinds of opportunities for think tanks and has bred different types of think tanks, too. But the demand for ideas from think tanks is there and has contributed to their development. Similarly, the differences in the number and strength of think tanks in Chile, South Africa, and India and their neighbours are partly explained by the institutional maturity of these countries.

Influence Takes Place and Changes over Time

Influence, too, cannot be described or measured over a prescribed and limited period of time, nor from a single perspective. Even specific policies are difficult to track and define around a particular moment of time. Policy debates develop over time and are driven by a complex of events, policy discourses, and interests that are difficult to predict or control for (Broadbent, 2012a).

Substantive influence then is closely associated with think tanks' capacity to maintain a course or develop a reputation on a particular issue or policy over the long term. The history of Chinese think tanks illustrates this point. Since the 1950s, they have been working hard at finding the right functional and structural balance and fit with their changing environment. A measure of their value cannot be limited to their influence on a single policy or event, but rather on their permanent presence in an otherwise monolithic State. What think tanks may be able to do today is, after all, in large part due to the sacrifices made and lessons learned in the past. In Chile, think tanks played a stabilisation role that directly affected the context in which policy was made; but this took place over a period of two decades. Only when think tanks and their researchers joined the government were they able to draw on research and ideas developed over that period of time. And even then, it would be impossible to credit these successes to think tanks alone without remembering the investments made in Chile's higher education during the two decades previous to the coup.

From a statist perspective, the policy discourse adopted by the State would dictate the access that certain think tanks have to the policy space. In Chile, the think tanks that were to become the brain of the democratic government in 1990 were dismissed by the military regime in the 1970s and 1980s. Similarly, the think tanks that appeared in the 1980s and 1990s to promote and then protect the economic liberalisation implemented by the military government had to wait until 2011, and a shift to the right, to regain influence. In Peru, think tanks have seen their influence rise and fall over the last few decades depending on the party and leaders in power. During the 1990s, the Government of Alberto Fujimori was open to technocrats willing to work within the limits of an authoritarian regime. Experts from private think tanks and consultancy-cum-think tanks dominated the discussion. After the return to democracy in 2000, access shifted in favour of experts from NGOs. Then access shifted in favour of relatively conservative economic policymaking at CIUP, during the government of Alan García (2006–2011), and towards GRADE and the researchers at Pontificia Universidad Católica del Perú, with Ollanta Humala's administration between 2011 and 2016. Since 2021, think tanks have lost ground and access altogether as a new regime led by new political elites has taken hold.

The Privatisation of Advice Can Limit the Different Roles That Think Tanks Can Play — Why Public Debate Is Important

Think tanks have a better chance of contributing in a variety of ways when there is a more plural community than when there are only one or two "experts" involved. Our discussions about the privatisation of policymaking — as described by Camou for Argentina, the history of think tanks in the United States, the adoption of the evidence-informed policy mantra particularly, by developing country governments and international development donors, the idea of evidence density, and Broadbent's focus on public policy debates as an entry point for research-informed evidence — support this lesson. Fewer plural systems, where a group within the elite or the State controls policymaking and the space available for think tanks and other knowledge producers and intermediaries, limit the opportunities for participation for those outside the prevailing discourse (Broadbent, 2012b).

They are likely to favour think tanks that legitimise their policy discourses and punish or dismiss those who challenge them.

In practice, there will be one or a few insiders and many more left out.

In more plural environments, the opportunities for several think tanks to play different roles increase. They can help start debates and discussions as in the case of JCTR's basic needs basket initiative, develop a collective memory or understanding of the lessons from the past as illustrated by Chile's think tanks' focus on reflecting on the changes that society underwent after the breakdown of democracy, map out and monitor the state of wellbeing of a country and the effect that policies to tackle poverty are having as SMERU has done, etc. Unfortunately, the trade-off is that as opportunities to play different roles increase, so does the difficulty in identifying and measuring exactly what they are.

The same is true for the existence of several levels of power or policy spaces. Highly centralised political systems focus think tanks' attention on a few institutions in the capital cities. More decentralised ones offer more opportunities for engagement, as well as for elites and local governments to seek to influence policy and encourage the formation of think tanks.

Policy issues controlled by the international community affect think tanks' access in the same manner. Broadbent's account of the GMO (Broadbent, 2012d) and urban decongestion (Broadbent, 2012c) debates in Zambia and Ghana, respectively, showed how the presence and absence of donors defined the role that research-based evidence and think tanks, in particular, played.

Having a Story to Tell

Overall, the relationship between think tanks and other political players is strengthened by the existence of common discourse and narratives. Knowing that there is a debate to join is not enough; one must also know how to join. Shared values and ideologies facilitate the translation of expert knowledge into more accessible information. Broadbent's study of African policy debates shows that think tanks play clearly identifiable roles when they are able to contribute to the development of a discourse — or to its critique. The same can be drawn from the accounts of the relationship between think tanks and the media by Ricardo Uceda (2011), Pablo Livszyc Natalia Romé (2011), and David Hojman (1997). Their studies helped to demonstrate that key policy players have their own agency and are active in communicating well-developed arguments.

This can also be found in the study of think tanks and politics in Latin America and East and Southeast Asia. The relationship with political parties or with the dominant political players in the different models of the developmental state is largely explained by a coincidence of values. The absence of this coincidence also explains the weak links that exist in certain contexts. For instance, a coincidence of values between think tanks and parties in Chile can be contrasted with the case of Peru, where the relationship between them is weak. Instead, researchers and politicians have opted to form and maintain technocratic networks or epistemic communities that act as informal meeting spaces.

Support to Think Tanks Matters

The 2024 State of the Sector Report (On Think Tanks, 2024) tackled this question head-on.

What explains think tanks' capacity to be more visible over a longer period of time, influence the identification of a problem or setting the agenda, sustaining an ideological alignment despite political instability, and developing coherent and convincing stories?

Think tanks' own organisational strength and resilience matters — and this is significantly affected by the form of funding they receive.

Their capacity to attract the right people and sustain their employment over a long period of time is greatly affected by their access to long-term and flexible funding.

So is think tanks' ability to develop and adapt their strategies in response to the opening or closing of windows of opportunity — rather than being limited by project-based contracting.

Core or programmatic funding is significantly easier to access in high-income countries than it is in low- and middle-income countries. In the latter, think tanks primarily rely on shorter-term project funding.

Sustaining an ideological alignment also demands funders that allow and encourage this. The existence of local political philanthropy makes this easier. Global research funders, which account for most low- and middle-income country think tank funding, are not privy to nor party to local political debates and cannot, therefore, support this type of approach.

Conclusions

The search for an appropriate measure of influence presents a number of challenges for think tanks and their supporters. In a way, this is the function that defines them and differentiates them from other organisations with which they share some characteristics. Thinking more carefully about it can certainly help them in their efforts to achieve it. There is much than can be done in the form of planning: understanding their political context and the research policy model that best describes their relationship with decision-makers, mapping out their main audience, and considering the best course of action to reach them and inform them of their ideas and proposals.

However, placing the same attention on assessing influence does not appear to be as straightforward as it might first seem. First of all, the prioritisation effort that takes place when planning what to do does not work well when trying to find out what happened. When think tanks adopt a perspective or assume a policy model to guide their work, they are making (or being made to make) a choice intended to help them work. They are in essence adopting decision-making models that can help make sense of a complex world and rely on simpler choices and actions. Without these models, they would not be able to determine the right balance between research and communications, or the breadth of their research agenda, the range of communication outputs and actions, the networks and alliances that best suit them, etc.

However, even when pursuing the paths that follow these simpler choices, think tanks still affect their environments by virtue of their own nature in more ways than one. Their relationships with other political players, the formal and informal agency of their researchers (many of whom are university lecturers and professors, public intellectuals, government officials, journalists, entrepreneurs, etc.), the turnover of staff, the development and maintenance of a collective memory, etc., provide several new avenues of influence that think tanks themselves my not have recognised ex ante as central to the strategy but are nonetheless present in everything they do. This influence, on people, processes, systems, and the environment, and by these other means, is no less important than that on identifiable policy decisions or choices.

If we add to this that influence is inter-temporal, arbitrary, mainly the consequence of the roles and agency of others, and by no means something that can be easily attributed to any one player in particular, then deciding where to look for it can present insurmountable challenges. A much more attractive option instead seems to be to focus on learning how to describe this complexity in terms that may be useful for the think tanks themselves.

The pluralist perspective offers an opportunity to develop a rich and nuanced account of how change happens, but it cannot guarantee an exact measure of think tanks' influence. Elitist and statist perspectives may get closer to this, but given that, in their view, think tanks are mostly pawns in someone else's game, then it is questionable whether their visibility or evidence that their research has been used is proof of their influence and not that they have been influenced to play that particular role. These perspectives, in any case, are certainly more useful when there is a clear demand for think tanks' outputs and their role is well recognised and valued.

Section 4

Chapter 11

Conclusions

As I warned at the beginning of this book, an inquiry into the role and nature of think tanks across the world poses a number of conceptual, methodological, and operational challenges. The first of these I attempted to deal with by reviewing the literature in search of definitions and descriptions of think tanks. I considered which were more relevant to think tanks in developing countries, which tend to be excluded from most studies, and put forward possible boundaries for the label. Most of these, however, remain hypothetical and are subject to interpretation and the particular characteristics of the context.

For the second challenge, I drew attention to the importance of the perspective of politics that we employ as a determining factor in the specific approach we take towards think tanks and their study. In my humble opinion, this is an important contribution to the literature as it shifts the focus away from a discussion on the details of think tanks' organisational characteristics to their contexts and even historical circumstances. This means that to study think tanks, we must study other political players and their relationships with think tanks.

Operationally, the challenge remains. The vast diversity of think tanks is such that any attempt to be thorough might be as elusive as the quest for an exact measure of influence. What can be provided, however, are examples or anecdotes of how some of the external and internal factors that describe think tank development play out in different contexts.

Delving too deeply into the internal workings of think tanks also proved to be particularly difficult. While organisations are by and large happy to be included in a study such as this, there is a clear concern that too much information may be divulged. That think tanks are political players, often dependent upon the patronage of powerful domestic and foreign forces, that their funding is for the most part precarious and currently facing changes, and that a culture of introspection is yet to develop make it difficult to take the organisational approach any further at this stage. I have therefore refrained from passing judgement throughout the account hitherto presented.

To facilitate this, learning imperatives rather than accountability demands, which are behind many attempts to assess and evaluate think tanks, should drive this type of inquiry. And as such, it may well be that on some occasions it will be better kept private or at least limited to being shared among the think tanks themselves.

To address this challenge, an effort should be made to fund and pursue comparative studies that bring the application of these different perspectives to more specific spaces of analysis. For example, rather than impossible like-for-like comparisons between think tanks in the same country or across countries, our efforts should seek to inquire about the similarities and differences among think tanks focused on similar policy debates. Is there anything unique to economic policy debates? Or to rural policy as opposed to urban policy? Are there any differences in the roles that think tanks play in these sectors across or within countries or regions? Are there particular national or regional characteristics that make some societies and their members friendlier to think tanks than others? Or, even more interestingly, why is it that different types of organisations act as think tanks in different countries and sectors?

Comparative studies should also attempt to link these perspectives together. Pluralism does not dismiss others. Rather, it creates a space for these to be deployed when necessary. While the pluralist perspective focuses our attention away from particular players and onto the policy discourses and debates, elitist and statist perspectives may help to describe the specific roles that these, and think tanks, have had on building and driving them. An organisational focus would then offer further evidence as to the particular functions played by think tanks in these processes.

Definitions, Functions, and Value

Throughout this book, I have made references to a number of organisations that share a common characteristic: They are all, by and large, interested in effecting change in issues of public interest. Some are more concerned than others with playing a leading role in public policy and are hence more likely to seek direct influencing strategies, while others prefer, for a number of reasons, a more tangential role and prioritise indirect approaches.

In my work, I often meet think tank funders who want to make the latter more like the former. Many organisations, too, among them some international development think tanks and consultancies, are devoting themselves to supporting this effort. I have myself contributed to it by developing methodologies and tools to support the incorporation of more systematically planned influencing strategies and monitoring and evaluation frameworks for think tanks and their funders to use.

What I think this book argues is that by focusing our attention on strengthening the communications components of think tanks (and other research programmes or initiatives), we are in fact affecting the very nature of these organisations. It is impossible to "insert" a new skill or competency without this having an effect on the whole. The challenge to define them stems from the large number of functional combinations and organisational options that are available for think tanks given the unique nature of their circumstances — influenced by different origins, missions, policy environments, staff composition, funding sources and mechanisms, etc. Among these, even a traditional focus on the production of academic knowledge, the indirect contribution to the policy debate by creating spaces and feeding them with ideas and people, and a low visibility approach are perfectly acceptable, and mostly appropriate, under certain circumstances.

The choice (of the mix of functions and structure) made by think tanks has been affected by a great number of internal and external factors that have left their mark on the organisation over the years. And whatever that mix is, we could assume that it works — for the centre would not be there otherwise.

Therefore, before attempting to change it, think tank leaders, their funders, and any interested service providers must be careful to ensure that they understand both the mix and its causes, and never assume that resistance to adopting new tactics or tools is due to irrationality, narrow self-interest, or simply based on ignorance.

In defining them, we must also be careful. Definitions and categories may be useful for scholars and funders but may be less relevant for think tanks themselves. While their staff may not know much about definitions of think tanks or the particular business models of their organisations, they are nonetheless able to recognise their unique nature and value. The definitions and categories reviewed at the start of this book unfortunately fail to describe them all in sufficient detail: They fail to convey the emotional aspects of these organisations and the manner in which their own staff view them and their contribution to them.

The functional approach to describing think tanks, combined with a recognition that these are defined in a self-identification process that both adopts and rejects the functions and organisational characteristics that others fulfil, presents as close an opportunity as seems possible. The functions, which can change over time and clearly depend on the roles played by other political players, help to describe the process by which think tanks contribute to society more effectively than normative definitions or the most detailed descriptions of their structure may ever do.

After all, what does size or thematic focus say about their value? Are larger think tanks more valuable than smaller ones? Is the nature of their funding an appropriate measure of their potential contribution? Is being not-for-profit intrinsically better than having parallel profitable interests or public responsibilities? Is academic research intrinsically better than empirical analysis? Questions about value are not easily answered by measurements of size and outputs.

Functions, on the other hand, highlight the agency of these organisations — and, together, they tell us about their purpose and motivations. It is this that points at their value and the specific contributions they can make to their societies.

In weak institutional contexts, in which the State, political parties, the private sector, the media, donors, and civil society more generally focus on short-term debates and objectives, think tanks can provide the *long-sightedness* that is necessary for development commitments. This purpose can be recognised in contexts as different as Chile, Zambia, and China.

Where parties were unable to develop programmatic proposals as in Chile, think tanks stepped in. Realising that the State was closed to the outside world and unable to build stable relations with others, Chinese leaders encouraged think tanks to do so instead. Think tanks can also fill the space left by an incompetent media by setting the agenda, informing and educating the public, legitimising and auditing policies, etc. They can compensate for poor academic standards by offering young graduates the opportunity to gain the skills and experience that they need to advance their careers. Think tanks often also act as advisors to politicians and policymakers, replacing formal advisory bodies, teams, or committees.

By adopting the functions of others, however, think tanks cannot pretend to replace them permanently as without them, they cannot expect to develop. Their own development is linked to the institutional development of all these other players. And it is precisely this symbiotic relationship that explains why think tanks may look and behave differently in different contexts (i.e., national or thematic). In essence, they are filling a gap left by others and as the gap changes — governments adopt new roles or capacities, academia is strengthened, the media becomes more or less active, etc. — think tanks must change, too.

Drivers

Both the absence of key public institutions (or their weakness) and their development can be said to constitute important drivers for the formation and development of think tanks. The history of think tanks in the United States begins with a response, on the part of the citizens, to the incapacity of the State to address public concerns. The same is true for the original Latin American think tanks: academic associations promoting the idea of an independent Peru in the later part of the 18th century. As states grew and became stronger, they developed their own analytical capacity and hence became more effective buyers or users of think thinks' ideas and services. The same is true for the relationship between think tanks and political parties, the media, and civil society more generally. Universities, too, have proved invaluable to the development of think tanks. Strong and unsuspecting hosts, such as the Church, have been the catalysts and drivers of these centres in many developing countries.

At the core of their history, however, is a powerful idea that has seen many incarnations: the role of science in policy. The metaphors described in this book have helped to keep the evidence-based policy mantra, whether we agree on its truth of not, alive and at the centre of the development of think tanks and policymaking. Even organisations set up with overtly ideological principles and objectives have used the claim to be evidence-based to their advantage. This idea has driven think tanks to take on different forms and shapes all around the world.

Other drivers can also be found. In certain circumstances, think tanks have provided an opportunity to germinate the ideas that serve as the basis for fundamental political liberalisations. This includes, of course, the independence of nations, but is more closely linked to the struggle to gain or regain democratic rule. In fact, it is not strange to find that their organisational arrangements and missions reflect this desire.

In other contexts, think tank have provided the basis for or have been used to promote certain economic reforms. The evidence-based policy discourse has offered an excellent vehicle to smuggle highly ideological policies under the cloak of technocratic choices. I never cease to be surprised at how polices such as cash transfers or universal access to certain basic services are presented as non-ideological solutions for developing countries when they are in fact highly contested issues in developed nations. In general, think tanks have promoted this idea as much as they have fed from it. International development agencies, keen to avoid any political meddling, prefer to direct their reform efforts to the funding and promotion of evidence that backs their views, and for this, think tanks, particularly contract think tanks, offer unbeatable advantages.

In any case, democracy can also be a driver for think tank formation but is certainly not a necessary one. Non-democratic regimes have the capacity to encourage their formation and development. The question is whether the regime, whatever its nature, values ideas and has a place for them in its politics.

There also appear to be important differences regarding the role that individuals or organisations play in the formation and development of think tanks. In some cases, we have seen think tanks imagined, set up, and managed by groups of individuals with a vision and a mission. These may be researchers or policymakers returning from work or study overseas, or even returning from working in the government, academia, or the private sector.

In other cases, this role has been taken on by funding bodies from within the public, private, or civil society sector, as well as foreign donors. Depending on their origin, they adopt different business models and core functions. The institutional origin also affects their governance which is in turn affected by their context.

By now, to say that context matters feels like a rather lazy assertion. What may be new is to argue that what matters is the agency of the dominating institutions in each context. Where political parties play a relatively dominant role as in Colombia and the United Kingdom, think tanks are bound to emerge from or around them and develop alongside them. Where it is the State or a single party as in China or Vietnam, think tanks will be naturally subject to their oversight. The corporate sector is bound to play a more powerful role in Japan and the United States, as it appears to be increasingly playing, at least in the debate on think tanks, in India. In Zambia and other Sub-Saharan African countries, foreign donors are still the driving forces with increasing participation of certain ministries.

This is important because it forces our attention and efforts slightly away from the think tanks to their contexts. But it is also telling of the non-passive nature of think tanks' own public or audiences. This is a finding that deserves special attention. The main difficulty in measuring think tanks' influence is that those that they wish to affect have an agency of their own and are therefore finally responsible for adopting or dismissing their advice. The elitist and statist perspectives award these other players a principal role in policymaking that relegates think tanks to a legitimating function — although other functions are also possible. The pluralist perspective allows think tanks to regain some protagonism, at least in smaller interactions, but this can be certainly nuanced by a careful assessment of the contributions that all other players make to the public debate.

The question over what motivates the development of think tank communities, specific think tanks, and those who work for them ought to focus our attention in the future.

On Influence

It is impossible to reflect on the issue of influence. The pressure to measure influence is rapidly building up for think tanks in developing countries — particularly for those benefiting from new funding opportunities in a context of increased concern for value for money and results. The literature on think tanks is clear on this issue: Measuring influence is an impossible and useless task.

The current approach, informed by accountability or marketing concerns, focuses on direct influence of policies: influencing processes that can be recorded and changes that can be attributed (at least partially) to the think tanks themselves and, more importantly even, to their funders. This organisation-centric perspective misses out on all the other valuable contributions made to society and to the environment in which policymaking takes place over an extended period of time.

Jeffrey Puryear said it best when he highlighted the psychological contribution made by Chilean think tanks. Their influence on policies (and there were many to report on) was overshadowed by their influence on the behaviour of political players in Chile. The emphasis that the

British Government in Zambia placed on the public economic debate for its future work with think tanks reflects this.

Assessing influence then ought to consider first the place that think tanks inhabit in their society and in relation to others — what roles or functions have they been called to play and why? All of these functions should be considered when addressing the question of influence. Furthermore, influence should not limit itself to measurable policy changes but also consider the invaluable contributions think tanks can make to the policymaking environment, the tools (and technologies) that policy actors employ, the skills and knowledge transferred to researchers, policy analysts, and policymakers, the creation and maintenance of spaces where knowledge and politics can come together, the quality of the public debate, etc.

It is indirectly and in the very long term that think tank influence is probably better understood, and so should not be dismissed by those looking for evidence of impact.

What Next?

Inevitably, some questions remain unanswered. Most notably, do we know what a think tank is? And more importantly, do we know what it is not? While the definition of think tanks is difficult to tackle at a global or even regional level, it is nonetheless important that an attempt is made at the national or local level. This definition is not intended to create a selective club but rather make the case for more and better targeted funding to all members of a policy community — including the various types of civil society organisations that, along with think tanks, make unique contributions. These definitions ought to pay attention to the boundaries between institutions and explore some of the nuances of the boundary nature of think tanks. They should allow for changes over time that reflect the fluidity of both think tanks and the label (Mendizabal, 2021).

In practical terms, the effort could be initially undertaken by either the local think tank community or by think tank funders, clarifying that these are not normative definitions but instead an attempt to encourage a conversation about the nature of this kind of organisation.

Much has been done already to map out and explain the relationships between think tanks and other political players in Latin America, and that effort has been replicated in Africa, South Asia, and South and Southeast

Asia. It would then be possible to compare across regions and maybe identify areas where one may learn from the other. Special attention needs to be provided to certain actors that have not featured in this study, for instance, the private sector, while the Church has made an unexpected appearance that ought to be explored. Both present an interesting focus of research.

An area of particular importance that remains understudied is funding. It was not possible, in this study, to assess exactly how much funding is allocated to think tanks as opposed to international development in general or research and civil society organisations in particular. Transparency efforts by donors do not appear to provide any sense that we will be able to measure how much funding is allocated to these organisations in developing countries, and much of it is shared across different types of funding mechanisms or vehicles. The same is true for information on funding from domestic sources, particularly from the private sector or individual philanthropists.

One approach we have opted for at On Think Tanks is to survey the think tanks themselves. The OTT State of the Sector Reports, published since 2021, offer insights into the sector from the perspective of the think tanks themselves.

As I have mentioned before, comparative research focused on different policy and thematic sectors would shed light on the working of different policy communities and how they affect think tanks. Comparisons among think tanks from the same countries or different regions within these communities may be more relevant and produce more useful findings than studies that treat them all equally and attempt to group think tanks together under traditions or waves.

A more in-depth study of individual think tanks' histories is necessary. It is possible that only *successful* ones will be willing to open up to scrutiny, but others should be encouraged to participate, too. These histories are important because they can describe and explain the forces that have promoted the formation and shaped the development of think tanks in developing countries. With similar attempts to describe their external environments, our understanding of think tanks may be complete.

Finally, a crucial albeit rather uncomfortable question could be tackled: Are think tanks necessary? In this study, a recurring factor explaining think tanks strength and influence is the degree of institutionalisation of all other types of organisations. A strong and professional civil service, a mature political system, a well-funded and dynamic academia, a stable and philanthropic private sector, and a professional and inquisitive media are sure indicators of a strong think tank community. But their weaknesses have also presented great opportunities for think tanks to develop.

It is in between these other players that the organisations that call themselves think tanks find it easier to operate, bringing people and spaces together, connecting problems with solutions, facilitating the movement of people as they build their careers criss-crossing different types of institutions, etc. Inevitably, the question arises: Are they just correcting a market failure? Can it be fixed?

At first sight, this appears to be the case. By their nature, they hold information about multiple players that would be too costly for any one of them to access. Policymakers do not have time to learn about all relevant academic researchers in all possible fields of study and deal with the day-to-day of political life. Similarly, academics cannot keep track of all policy debates and processes in search of the right windows of opportunity while conducting sound and robust research. As think tanks are free from these other more pressing responsibilities, they can.

The same is true for the relationship with the media. The account of public intellectuals showed that think tanks can provide expert commentary on a vast range of subjects that the media alone would never be able to master. Think tanks are also able to channel funds from philanthropists into academia (by their use of academic researchers in their projects) and parties (offering advice and even staff) which would be impossible to provide directly.

The concept of "windows" presented in the case of Chinese think tanks illustrates this at the international level. Think tanks provide an intermediary service between different political and academic cultures, even between societies. More worrying still, their value is in the creation and maintenance of connecting or facilitating spaces.

Will the rapidly developing world of AI offer hitherto unimagined opportunities to create direct connections between citizens, funders, policymakers, researchers, the media, and politicians in a way that will see think tanks, at least as we know them, become a thing of the past (Mendizabal, 2024)?

Bibliography

Abelson, D. E. (2006). *A Capitol Idea: Think Tanks and US Foreign Policy.* Montreal: McGill-Queen's University Press.

Abelson, D. E. (2009). *Do Think Tanks Matter? Assessing the Impact of Public Policy Institutes.* Quebec City: McGill-Queen's University Press.

Adusei, K. O. and Yeo, S. (2010). Building Capacity for Policy Research in Southern Africa. Draft, The World Bank, Development Economics Vice-Presidency.

Alansi, A. M. (2021). *Think Tanks in Saudi Arabia and their Role in Guiding General Policy.* Riyadh: Center for Research and Intercommunication Knowledge.

Alvarado, L. (2023, August 29). *Watchdog think tanks.* Retrieved from OTT Talks: https://open.spotify.com/episode/177LmmhKMHK7xwcMUqJKfn.

Alvarado, L. (2024, June 4). The relationship between think tanks and their communities. *On Think Tanks.*

Ambrose, N. (2005). Global Philanthropy: Emerging Issues and Strategic Planning. Council on Foundations.

American Association for the Promotion of Social Science. (1866). Constitution, Address, and List of Members of the American Assocaition for the Promotion of Social Science, with Questions Proposed for Discussion: To Which Are Added Munites of the Transactions of the Association. Boston: Wright and Poter.

Baertl, A. (2018, March). De-constructing credibility Factors that affect a think tank's credibility. *OTT Working Paper.*

Bai, M. (2008). *The Argument: Billionaires, Bloggers, and the Battle to Remake Democratic Politics.* Penguin Group.

Bajpai, K. (2010, April 3). Think tanks in India's democracy. Retrieved July 10, 2011, from *The Times of India*: http://articles.timesofindia.indiatimes.com/2010-04-03/edit-page/28137957_1_tanks-policy-public-debate.

Baru, S. (2010, August 9). IMAGINDIA. Retrieved July 10, 2011, from Indian Minds, Foreign Funds: http://www.imagindia.org/debate_aug10.html.

Belletini, O. (2007). El papel de los centros de investigacion de politica publica en las reformas publicas implementadas en America Latina. In *Think Tanks y politicas publics en Latinoamerica: Dinamicas globales y realidades regionales.* Buenos Aires: IDRC/Konrad Adenauer Stiftung/ Prometeo libros.

Bellettini, O. (2011, August 8). On Think Tanks. Retrieved September 10, 2011, from Think Tanks and politics/Think tanks y la política: http://onthinktanks. org/2011/08/08/think-tanks-and-politics-think-tanks-y-la-politica/.

Bellettini, O. and Carrión, M. (2009). Partidos políticos y think tanks en el Ecuador. In Mendizabal, E. and Sample, K. (Eds.), *Dime a quién escuchas... think tanks y partidos políticos en América Latina.* Lima: IDEA/ODI.

Bery, S. (2010, December 14). *Business Standard.* Retrieved July 9, 2011, from Taking think tanks seriously: India's think tanks need to diversify their support to enhance their contribution: http://www.business-standard.com/india/ news/suman-bery-taking-think-tanks-seriously/418065/.

Bhagwati, J. (2010, August 12). *The Times of India.* Retrieved July 9, 2011, from Expanding India's Expertise: http://articles.timesofindia.indiatimes. com/2010-08-21/edit-page/28281534_1_foreign-funding-indian-ngo-csr#ixzz17ooVkA7X.

Booth, D. (2011, April). Working with the grain and swimming against the tide: Barriers to uptake of research findings on governance and public services in low-income Africa. *Working Paper.* Africa Power and Politics.

Botto, M. (2011). Think tanks en América Latina: radiografía comparada de un nuevo actor politico. In Correa, N. and Mendizabal, E. (Eds.), *Vínculos entre conocimiento y política: el rol de la investigación en el debate public en América Latina.* Lima: Universidad del Pacifico/CIES.

Bradford, N. (1998). *Commissioning Ideas: Canadian National Policy Innovation in Comparative Perspectives.* Don Mills: Oxford University Press.

Braml, J. (2004). *Think Tanks versus "Denkfabriken"? U.S. and German Policy Research Institutes' Coping with and Influencing Their Environments.* Baden-Baden: Nomos Verlagsgesellshaft.

Braun, M., Chudnovsky, M., Ducote, N., and Weyrauch, V. (2007). Lejos de "Thinktanklandia": los institutos de investigacion de politicas en los paises en desarrollo. In Garce, A. and Una, G. (Eds.), *Think Tanks y politicas publicas en Latinoamerica: Dinamicas globales y realidades regionales.* Buenos Aires: IDRC/Konrad Adenauer Stiftung/Prometeo Libros.

Broadbent, E. (2012a). Policy debates in Africa: So what, and what now? *First International Evidence Informed Policy Conference.* Ile.

Broadbent, E. (2012b). *Politics of Research-Based Evidence in African Policy Debates: Synthesis of Case Study Findings.* London: EBPDN/Mwananchi.

Broadbent, E. (2012c). *Research-Based Evidence in African Policy Debates. Case Study 1: Decongestion in Accra, Ghana.* London: EBPDN/ Mwananchi.

Broadbent, E. (2012d). *Research-Based Evidence in African Policy Debates. Case Study 3: The Contemporary Debate on Genetically Modified Organisms in Zambia.* London: ebpdn/Mwananchi.

Camou, A. (2007). Think tanks en Argentina: sobreviviendo a la tensión entre la participación y la permanencia. In Uña, G. and Garcé, A. (Eds.), *Think tanks y políticas públicas en Latinoamérica : dinámicas globales y realidades regionales.* Buenos Aires: Prometeo Libros.

Camp, R. A. (1998). Technocracy a la Mexicana: Antecedent to Democracy? In Centeno, M. A. and Silva, P. (Eds.), *The Politics of Expertise in Latin America.* Basingstoke: MacMillan Press.

Carden, F. (2009). *Knowledge to Policy: Making the Most of Development Research.* New Delhi: Sage Publications/IDRC.

Centeno, M. A. and Silva, P. (1998). The politics of expertise in Latin America: Introduction. In Centeno, M. A. and Silva, P. (Eds.), *The Politics of Expertise in Latin America.* Basingstoke: MacMillan Press.

Chen, S.-C. J. (2007, October 19). Forbes.com. Retrieved December 20, 2011, from When Asia's Millionaires Splurge, They Go Big: http://www.forbes.com/2007/10/19/asia-rich-report-face-markets-cx_jc_1019autofacescan01.html.

Clarke, J., Mendizabal, E., Leturque, H., Walford, V., and Pearson, M. (2009). *DFID Influencing in the Health Sector: A Preliminary Assessment of Cost Effectiveness.* London: ODI.

Clifton, J. (2010, June 09). Strategic promotion of ageing capacity building. Retrieved February 14, 2011, from http://www.sparc.ac.uk/workshops/2010-06-09-research-into-policy/pdf/Jonathan.pdf.

Cociña, M. and Toro, S. (2009). Los think tanks y su rol en la arena política chilena. In Mendizabal, E. and Sample, K. (Eds.), *Dime a quién escuchas... think tanks y partidos políticos en América Latina.* Lima: IDEA/ODI.

Correa, N. and Mendizabal, E. (Eds.). (2011). *Vínculos entre conocimiento y política: el rol de la investigación en el debate público en América Latina.* Lima: Universidad del Pacifico/CIES.

Court, J. and Young, J. (2003). Bridging research and policy: Insights from 50 case studies. *Working Paper, 213.*

Da Costa, P. K. (2011). *Rule of Experts?: Decomposing Agency and Agendas in Africa's Development Regime.* PhD thesis. School of Oriental and African Studies.

Datta, A. (2021). Rethinking organisational development: an OTT series. *On Think Tanks.* Retrieved from https://onthinktanks.org/series/rethinking-organisational-development-an-ott-series/.

deGrassi, A. (2007, November). Linking research and policy: The case of Ghana's rice trade policy. *Background Paper No. GSSP 0010.*

Dye, T. (1976). *Who's Running America: Institutional Leadership in the United States.* New Jersey: Prentice Hall.

Echt, L. (2019, August). Partisan think tanks: between knowledge and politics The case of Pensar Foundation and PRO party in Argentina. *On Think Tanks Working Paper.*

Economist Intelligence Unit. (2011). Something's gotta give: The state of philanthropy in Asia. *The Economist.*

Fowler, W. (1997). Introduction: Stressing the importance of ideological discourse. In Fowler, W. (Ed.), *Ideologues and Ideologies in Latin America.* Westport: Greenwood Press.

Freeman, B. and Cleveland-Stout, N. (2025). *The Case for Think Tank Transparency.* Retrieved from Quincy Institute: https://quincyinst.org/2025/01/16/the-case-for-think-tank-transparency/.

Fundación PROhumana and Fundación Ford. (2003). La filantropía en América Latina: los desafíos de las fundaciones donantes en la construcción de capital humano y justicia social. *Seminario Internacional Fundación PROhumana y Fundación Ford.* Santiago de Chile.

Garcé, A. (2009). Estudio marco. In Mendizabal, E. and Sample, K. (Eds.), *Dime a quién escuchas... think tanks y partidos politicos en América Latina.* Lima: IDEA/ODI.

Garce, A. (2018, January). Political Knowledge Regimes and policy change in Chile and Uruguay. *On Think Tanks Working Paper.*

Garce, A. and Una, G. (Eds.). (2007). *Think Tanks y politicas publicas en Latinoamerica: Dinamicas globales y realidades regionales.* Buenos Aires: IDRC/Konrad Adenauer Stiftung/Prometeo Libros.

Gill, B. and Mulvenon, J. (2002, September). Chinese military-related think tanks and research institutions. *The China Quarterly, 171*, 617–624.

Glaser, B. S. and Saunders, P. C. (2002, September). Chinese civilian foreign policy research institutes: Evolving roles and increasing influence. *The China Quarterly, 171*, 597–616.

Gonzalez Gutierrez, C. (2006). Introduccion: El paper de los gobiernos. In Gonzalez Gutierrez, C. (Ed.), *Relaciones Estado-diaspora: la perspectiva de America Latina y el Caribe.* Mexico DF: Secretaria de Relaciones Exteriores.

Go To Think Tank Index. (2012). Retrieved January 25, 2012, from Think Tanks and Civil Societies Program: http://www.gotothinktank.com/.

Groseclose, T. and Milyo, J. (2005, November). A measure of media bias. *The Quarterly Journal of Economics (O. U. Press Edition), 120*(4), 1191–1237.

Gusterson, H. (2009). *The Sixth Branch: Think Tanks as Auditors.* New York: Social Science Research Council.

Guttman, D. and Willner, B. (1976). *The Sadow Government: The Government's Multi-Billion-Dollar Giveaway of its Decision-Making Powers to Private Management Consultants, "Experts," and Think Tanks.* New York: Pantheon Books.

Harvard Law Review. (2002, March). The political activity of think tanks: The case for mandatory contributor disclosure. *Harvard Law Review, 115*(5), 1502–1524.

Hayter, E. and Makokha, R. (2024, July 1). Government research funding reform in West Africa: what to watch. *On Think Tanks.*

Hojman, D. E. (1997). El Mercurio's editorial page ("La Semana Economica") and neoliberal policy making in today's Chile. In Fowler, W. (Ed.), *Ideologues and Ideologies in Latin America.* Westport: Greenwood Press.

Hoppe, R. (2010). From "knowledge use" towards "boundary work": Sketch of an emerging new agenda for inquiry into science-policy interaction. In in't Veld, R. J. (Ed.), *Knowledge Democracy: Consequences for Science, Politics, and Media.* Heidelberg: Springer.

in't Veld, R. J. (2010). Towards knowledge democracy. In in't Veld, R. J. (Ed.), *Knowledge Democracy: Consequences for Science, Politics, and Media.* Heildelberd: Springer.

Jones, H. (2011, February). A guide to monitoring and evaluating policy impact. *ODI Background Note.* London: ODI.

Jones, N. and Young, J. (2007). *Setting The Scene: Situating DFID's Research Funding Policy and Practice in an International Comparative Perspective.* London: Overseas Development Institute, Research and Policy in Development.

Kimenyi, M. S. and Datta, A. (2011). *Thinking Politics: How the Political Lanscape Has Informed the Development of Think Tanks in Sub-Saharan Africa.* London: Overseas Development Institute (ODI), Research and Policy in Development.

Klein, G. (2001). *Sources of Power: How People Make Decisions.* Cambridge: The MIT Press.

Klotz, A. (2000, October). Migration after apartheid: Deracialising South African foreign policy. *Third World Quarterly, 21*(5), 831–847.

Kuljian, C. L. (2005). *Philanthropy and Equity: The Case of South Africa.* Global Equity Initiative. Harvard University.

Kwaku Ohemeng, F. L. (2005, September). Getting the state right: Think tanks and the dissemination of new public management ideas in Ghana. *The Journal of Modern African Studies, 43*(3), 443–465.

Lardone, M. and Roggero, M. (2011). El rol del estado en el financiamiento de la investigación sobre políticas públicas en América Latina. In Correa, N. and Mendizabal, E. (Eds.), *Vínculos entre conocimiento y política: el rol de la investigación en el debate público en América Latina.* Lima: Universidad del Pacifico/CIES.

Li, C. (2009). China's new think tanks: Where officials, entrepreneurs, and scholars interact. *China Leadership Monitor, 29*, 1–21.

Lipton, E., Brooke , W., and Nicholas , C. (2014, September 6). Foreign powers buy influence at think tanks. *The New York Times.*

Lipton, E., Williams, B., and Confessore, N. (2024, September 6). Foreign Powers Buy Influence at Think Tanks. *The New York Times.*

Livszyc, P. and Romé, N. (2011). Medios de comunicación y uso de la investigación en políticas públicas en América Latina. In Correa, N. and Mendizabal, E. (Eds.), *Vínculos entre conocimiento y política: el rol de la investigación en el debate público en América Latina.* Lima: Universidad del Pacifico/CIES.

Loayza, R. (2011). Think tanks: los medios de poder en la Bolivia de Evo Morales. In Correa, N. and Mendizabal, E. (Eds.), *Vínculos entre conocimiento y política: el rol de la investigación en el debate público en América Latina.* Lima: Universidad del Pacifico/CIES.

Londoño, J. F. (2009). Partidos politicos y think tanks en Colombia. In Mendizabal, E. and Sample, K. (Eds.), *Dime a quién escuchas... think tanks y partidos politicos en América Latina.* Lima: IDEA/ODI.

Lowry, R. C. (1999, August). Foundation patronage toward citizen groups and think tanks: Who get grants? *The Journal of Politics, 61*(3), 758–776.

Lusthaus, C., Adrien, M.-H., Anderson, G., Cardon, F., and Plino Montalvan, G. (2002). *Organizational Assessment: A Framework for Improving Performance.* Ottawa/Washington D.C.: Inter-American Development Bank and International Development Research Centre.

Luther, J. (2006, March). I-searching in context: Thinking critically about the research unit. *The English Journal (N. C. English Edition), 95*(4), 68–74.

McGann, J. and Weaver, K. (Eds.). (2002). *Think Tanks and Civil Societies: Catalysts for Ideas and Action.* Transaction Publishers.

McNutt, K. and Marchildon, G. (2009, June). Think tanks and the web: Measuring visibility and influence. *Canadian Public Policy/Analyse de Politiques, 35*(2), 219–236.

Medvetz, T. (2008). *Think Tanks as an Emergent Field.* New York: Social Science Research Council.

Mendizabal, E. (2006a). Building effective research policy networks: Linking function and form. *ODI Working Paper, 276.*

Mendizabal, E. (2006b). Understanding networks: The form and function of research policy networks. *ODI Working Paper, 271.*

Mendizabal, E. (2011a, November 20). A new idea: Do not fund think tanks. Retrieved December 15, 2011, from onthinktanks.org: http://onthinktanks.org/2011/11/20/a-new-idea-do-not-fund-think-tanks/.

Mendizabal, E. (2011b, March 9). Different ways to define and describe think tanks. Retrieved July 10, 2011, from On Think Tanks: http://onthinktanks.org/2011/03/09/different-ways-to-define-and-describe-think-tanks/.

Mendizabal, E. (2011c, July 20). Never mind the gap: On how there is no gap between research and policy and on a new theory (Part 3 of 3). Retrieved December 10, 2011, from onthinktanks.org: http://onthinktanks.org/2011/07/20/never-mind-the-gap-on-how-there-is-no-gap-between-research-and-policy-and-on-a-new-theory-part-3-of-3/.

Mendizabal, E. (2012, November 23). Chinese think tanks: From windows to super highways. *On Think Tanks*.

Mendizabal, E. (2016a, February 2). The New Philanthropists: Rohini Nilekani, Founder and Chairperson of Arghyam. *On Think Tanks*.

Mendizabal, E. (2016b, April 27). Think Tanks in China: A Golden Age? *On Think Tanks*.

Mendizabal, E. (2019, February 2). The flaws in the ranking. *On Think Tanks*.

Mendizabal, E. (2021, August 4). The future of think tanks. *On Think Tanks*.

Mendizabal, E. (2022, September 5). Playing the long game: Politics, elite bargaining, and change over 20 years in Peru. *On Think Tanks*.

Mendizabal, E. (2024a). An Insider's Perspective into Equitable Think Tank Partnerships: Time for a Real Change. *On Think Tanks*.

Mendizabal, E. (2024b). Not influential but useful: Rethinking how we assess and support think tanks. *On Think Tanks*.

Mendizabal, E. (2024c). The promise and perils of AI in shaping tomorrow's think tanks, policymakers and foundations. *On Think Tanks*.

Mendizabal, E. (2025, February 27). As the world turns authoritarian: Strategies for think tanks and their funders. *On Think Tanks*.

Mendizabal, E. and Hearn, S. (2011). Not everything that connects is a network. *ODI Background Note*.

Mendizabal, E. and Jones, H. (2010). *Strengthening Learning from Research and Evaluation: Going with the Grain.* London: ODI.

Mendizabal, E. and Sample, K. (Eds.). (2009). *Díme a quien escuchas... Think tanks y partidos políticos en América Latina.* Lima: ODI/IDEA International.

Mendizabal, E. and Yeo, S. (2010, June 30). The Broker. Retrieved December 15, 2011, from The virtues of virtuality: from think tank to think net: http://www.thebrokeronline.eu/Articles/The-virtues-of-virtuality.

Mendizabal, E., Datta, A., and Jones, N. (2010). Think tanks and the rise of the knowledge economy. In Garcé, A. and Uña, E. (Eds.), *Think Tanks and Public Policies in Latin America.* Fundación Siena/CIPPEC.

Mendizabal, E., Datta, A., and Young, J. (2011). Developing capacities for better research uptake: The experience of ODI's research and policy in development programme. *ODI Background Notes (O. D. Institute Edition).*

Michael, S. (2025, January 3). 'Dark Money' is tainting Washington think tanks. A new report shows it's worse than you think. *Politico*.

Mkandawire, T. (2000). Non-organic intellectuals and 'learning' in policy-making Africa. *Learning in Development Co-operation*.

Nachiappan, K., Mendizabal, E., and Datta, A. (2010). *Think Tanks in East and Southeast Asia.* London: ODI.

Naughton, B. (2002, September). China's economic think tanks: Their changing role in the 1990s. *The China Quarterly*, 625–635.

On Think Tanks. (2024). *The State of the Sector Report 2024: Resilience and Impact in a Politically Shifting World.* On Think Tanks.

Overseas Development Institute. (2004, October). Bridging research and policy in international development an analytical and practical framework. *Briefing Paper.*

Posner, R. A. (2001). *Public Intellectuals: A Study of Decline.* Cambridge: Harvard University Press.

Puryear, J. M. (1994). *Thinking Politics: Intellectuals and Democracy in Chile, 1973–1988.* Baltimore: The Johns Hopkins University Press.

Ralphs, G. (2016, June 14). Think tank business models: The business of academia and politics. *On Think Tanks.*

Raucher, A. (1978, Autumn). The first foreign affairs think tanks. *American Quarterly, 30*(4), 493–513.

Reinicke, W. H. (1996). *Tugging at the Sleeves of Politicians: Think Tanks -American Experiences and German Perspectives.* Gütersloh: Bertelsmann Foundation Publishers.

Ricci, D. M. (1993). *The Transformation of American Politics: The New Washington and the Rise of Think Tanks.* New Haven: Yale University Press.

Rich, A. (2006). *Think Tanks, Public Policy and the Politics of Expertise.* New York: Cambridge University Press.

Ruiz Rodriguez, L. M. (2006). La Coherencia Programatica en los Partidos Politicos. In Alcantara Saez, M. (Ed.), *Politicos y politica en America Latina.* Madrid: Fundacion Carolina/Siglo XXI.

Russell, W. (1982a, October 16). Letter from Westminster: Think tank's report shelved, but not ditched. *British Medical Journal (Clinical Research Edition), 285*(634), 1133.

Russell, W. (1982b, October 16). Think Tank's report shelved, but not ditched. *British Medical Journal, 285*, 1133.

Sanghi, A. (2010, November 24). Quick take -As it happens. Retrieved July 9, 2011, from Collusion or collaboration? The Think Tank Initiative: http://quicktake.wordpress.com/2010/11/24/collusion-or-collaboration-the-think-tank-initiative/.

Scartascini, C. (2008). Who's who in the PMP: An overview of actors, incentives, and the roles they play. In Stein, E. and Tommasi, M. (Eds.), *Policymaking in Latin America: How Politics Shapes Policies.* Washington DC: Inter-American Development Bank/David Rockefeller Center for Latin American Studies.

Scott, N. (2011, September 12). Digital disruption Responding to digital disruption of traditional communications: Three planks to ODI's digital strategy. Retrieved December 14, 2011, from onthinktanks.org: http://onthinktanks. org/2011/09/12/responding-digital-disruption-traditional-communications-odi-strategy/.

Shambaugh, D. (2002, September). China's international relations think tanks: Evolving structures and process. *The China Quarterly, 171*, 575–596.

Sheth, A. and Singhal, M. (2011). *India Philanthropy Report 2011.* Bain & Company.

Siavelis, P. and Morgenstern, S. (Eds.). (2008). *Pathways to Power: Political Recruitment and Candidate Selection in Latin America.* University Park: The Pennsylvania State University Press.

Smith, J. A. (1991). *The Idea Brokers: Think Tanks and the Rise of the New Policy Elite.* New York: The Free Press.

Snowdon, P. (2010). *Back from the Brink: The Extraordinary Fall and Rise of the Conservative Party.* London: Harper Press.

Spiller, P. T., Stein, E., and Tommasi, M. (2008). Political institutions, policy-making, and policy: An introduction. In Stein, E. and Tommasi, M. (Eds.), *Policymaking in Latin America: How Politics Shapes Policies.* Washington DC: Inter-American Development Bank/David Rockefeller CEnter for Latin American Studies.

Srivastava, J. (2012). *Think Tanks in South Asia: Analysing the Knowledge-Power Interface.* London: Overseas Development Institute.

Stein, E. and Tommasi, M. (Eds.). (2008). *Policymaking in Latin America: How Politics Shapes Policies.* Washington DC: Inter-American Development Bank/David Rockefeller Center for Latin American Studies.

Stojanovic-Gajic, S. (2025, January 22). *Running a Think Tank in a Changing Europe: Adapting to New Realities in the Western Balkans.* Retrieved from OTT Talks: https://onthinktanks.org/resource/running-a-think-tank-in-a-changing-europe-adapting-to-new-realities-in-the-western-balkans/.

Stone, D. (2008). Global public policy, transnational policy communities and their networks. *Policy Studies Journal, 36*(10), 19–38.

Stone, D. and Denham, A. (2004). *Think Tank Traditions: Policy Research and the Politics of Ideas.* Manchester: Manchester University Press.

Struyk, R. J. (2006). *Managing Think Tanks: Practical Guidance for Maturing Organizations* (2nd edn.). Budapest: OSI/LGI, The Urban Institute.

Struyk, R. J., Damon, M., and Haddaway, S. R. (2011). Running ahead: Evaluating policy research organization mentoring programs. *American Journal of Evaluation.*

Sumarto, S. (2011). The SMERU Research Institute: History and Lessons Learned. Final Draft Report, Australia-Indonesia Partnership, Revitalising Indonesia's Knowledge Sector for Development, Jakarta.

Tanaka, M., Barrenechea, R., and Morel, J. (2011). La relación entre investigacion y políticas públicas en América Latina: un análisis exploratorio.
In Correa, N. and Mendizabal, E. (Eds.), *Vínculos entre conocimiento y
política: el rol de la investigación en el debate público en América Latina.*
Lima: Universidad del Pacifico/CIES.

Tanaka, M., Vera, S., and Barrenechea, R. (2009). Think tanks y partidos politicos
en el Perú: precariedad institutional y redes informales. In Mendizabal, E.
and Sample, K. (Eds.), *Dime a quién escuchas... think tanks y partidos
politicos en América Latina.* Lima: IDEA/ODI.

Tanner, J. (2012, June 18). Social media and think tanks: Lessons from
London Thinks. Retrieved June 30, 2012, from onthinktanks.org: http://
onthinktanks.org/2012/06/18/social-media-and-think-tanks-lessons-from-
london-thinks/.

Tanner, M. S. (2002, September). Changing windows on a changing China: The
evolving "think tank" system and the case of the public security sector. *The
China Quarterly, 171*, 559–574.

The Economist. (2011, November 16). *The Economist.* Retrieved December 15,
2011, from Have PhD, will govern: http://www.economist.com/blogs/news-
book/2011/11/technocrats-and-democracy.

Toranzo, C. (2009). Partidos políticos y think tanks en Bolivia. In Mendizabal, E.
and Sample, K. (Eds.), *Dime a quién escuchas... think tanks y partidos
políticos en América Latina.* Lima: IDEA/ODI.

Tutchings, T. R. (1979). *Rhetoric and Reality : Presidential Commissions and the
Maing of Public Policy.* Boulder: Westview Press.

Uceda, R. (2011). Una extraña pareja. La relación entre los medios de comunicación y los centros de investigacion en políticas públicas. In Correa, N. and
Mendizabal, E. (Eds.), *Vínculos entre conocimiento y política: el rol de la
investigación en el debate público en América Latina.* Lima: Universidad del
Pacifico/CIES.

Walsh, B. (2006, September 4). *Times Magazine World.* Retrieved December 20,
2011, from Learning the Art of Giving: http://www.time.com/time/magazine/
article/0,9171,1531426,00.html#ixzz1hjMK2vYt.

Warden, J. (1988, February 13). Letter from Westminster: The think tanks roll.
British Medical Journal (Clinical Research Edition), 296(6620), 510.

Weaver, R. K. (1989, September). The changing world of think tanks. *PS:
Political Science and Politics, 22*(3), 563–578.

Weidenbaum, M. (2009). *The Competition of Ideas: The World of the Washington
Think Tanks.* New Jersey: Transaction Publishers.

Weyland, K. (2006). *Bounded Rationality and Policy Diffusion: Social Sector
Reform in Latin America.* Princeton: Princeton University Press.

Wiess, C. (1977). Research for policy's sake: The enlightenment function of
social research. *Policy Analysis, 3*(4), 531–545.

Yeo, S. and Portes, R. (2001). *'THINK-NET': The CEPR Model of a Research Network*. CEPR.

Yeoh, T. (2024, June 4). Building bridges across communities and bottom up accountability. *On Think Tanks*.

Zufeng, Z. (2009, March-April). The influence of think tanks in the Chinese policy process: Different W. *Asian Survey, 49*(2), 333–357.

Index

www.ingramcontent.com/pod-product-compliance
Lightning Source LLC
Chambersburg PA
CBHW050637190326
41458CB00008B/2312